I TOO HAD A DREAM

Born in Calicut, Kerala, **Dr Verghese Kurien** (1921–2012) graduated in science and engineering from Madras University and Michigan State University, USA, respectively.

He began his career in dairying at the government's creamery in Anand, Gujarat, later joining the Kaira District Cooperative Milk Producers' Union Limited (now Amul). As Chairman of the National Dairy Development Board, he implemented 'Operation Flood'.

Recipient of countless awards, including the Ramon Magsaysay Award (1963), Wateler Peace Prize (1986), World Food Prize (1989), Padma Shri (1965), Padma Bhushan (1966), and Padma Vibhushan (1999), Dr Kurien was also Chairman of the Institute of Rural Management, Anand, Chairman of the Gujarat Cooperative Milk Marketing Federation, and Chairman of the National Cooperative Dairy Federation of India.

Gouri Salvi is a Mumbai-based freelance journalist. She has worked with *Onlooker* and *Sunday* magazines and with the *Women's Feature Service*. She has written on development and gender issues, has co-edited *Beijing!* a book on the UN's Fourth World Conference on Women, and edited *Development Retold: Voices From the Field*, a book on the Indian Cooperative Union.

OTHER LOTUS TITLES

Anil Dharker	Icons: Men & Women Who Shaped Today's India
Aitzaz Ahsan	The Indus Saga: The Making of Pakistan
Ajay Mansingh	Firaq Gorakhpuri: The Poet of Pain & Ecstasy
Alam Srinivas & T.R. Vivek	IPL: The Inside Story
Alam Srinivas	Women of Vision: Nine Business Leaders in Conversation
Amarinder Singh	The Last Sunset: The Rise & Fall of the Lahore Durbar
Aruna Roy	The RTI Story: Power to the People
Ashis Ray	Laid to Rest: The Controversy of Subhas Chandra Bose's Death
Bertil Falk	Feroze: The Forgotten Gandhi
Harinder Baweja (Ed.)	26/11 Mumbai Attacked
Harinder Baweja	A Soldier's Diary: Kargil – The Inside Story
Ian H. Magedera	Indian Videshinis: European Women in India
Kunal Purandare	Ramakant Achrekar: A Biography
Lucy Peck	Agra: The Architectural Heritage
Lucy Peck	Delhi a Thousand Years of Building: An INTACH-Roli Guide
Madan Gopal	My Life and Times: Munshi Premchand
M.J. Akbar	Blood Brothers: A Family Saga
Maj. Gen. Ian Cardozo	Param Vir: Our Heroes in Battle
Maj. Gen. Ian Cardozo	The Sinking of INS Khukri: What Happened in 1971
Madhu Trehan	Tehelka as Metaphor
Manish Pachouly	The Sheena Bora Case
Moin Mir	Surat: Fall of A Port Rise of A Prince Defeat of the East India Company in the House Of Commons
Monisha Rajesh	Around India in 80 Trains
Noorul Hasan	Meena Kumari: The Poet
Prateep K. Lahiri	A Tide in the Affairs of Men: A Public Servant Remembers
Rajika Bhandari	The Raj on the Move: Story of the Dak Bungalow
Ralph Russell	The Famous Ghalib: The Sound of my Moving Pen
Rahul Bedi	The Last Word: Obituaries of 100 Indian who Led Unusual Lives
R.V. Smith	Delhi: Unknown Tales of a City
Salman Akhtar	The Book of Emotions
Sharmishta Gooptu	Bengali Cinema: An Other Nation
Shrabani Basu	Spy Princess: The Life of Noor Inayat Khan
Shahrayar Khan	Bhopal Connections: Vignettes of Royal Rule
Shantanu Guha Ray	Mahi: The Story Of India's Most Successful Captain
S. Hussain Zaidi	Dongri to Dubai
Sunil Raman & Rohit Aggarwal	Delhi Durbar: 1911 The Complete Story
Thomas Weber	Going Native: Gandhi's Relationship with Western Women
Thomas Weber	Gandhi at First Sight
Vaibhav Purandare	Sachin Tendulkar: A definitive biography
Vappala Balachandran	A Life In Shadow: The Secret Story of ACN Nambiar – A Forgotten Anti-Colonial Warrior
Vijayan Bala	The Complete Indian Sports Quiz Book
Vir Sanghvi	Men of Steel: India's Business Leaders in Candid Conversation

FORTHCOMING TITLES

Raju Santhanam	The Untold Story of Christian Michel and Agusta Westland

I TOO HAD A DREAM

VERGHESE KURIEN

as told to
GOURI SALVI

FOREWORD BY RATAN TATA

Lotus Collection

© Verghese Kurien, 2005

All rights reserved. No part of this publication
may be reproduced or transmitted, in any form
or by any means, without the prior permission of
the publisher.

First published in India in 2005
This paperback edition published in 2007
Seventeenth impression, 2019
The Lotus Collection
An imprint of
Roli Books Pvt. Ltd.
M-75, Greater Kailash II Market, New Delhi 110 048
Phone: ++91 (011) 4068 2000
E-mail: info@rolibooks.com
www.rolibooks.com
Also at Bengaluru, Chennai & Mumbai

Cover design: Sneha Pamneja
Cover picture: Getty Images

ISBN: 978-81-7436-407-4

Typeset in Photina MT by Roli Books Pvt. Ltd.
Printed at Sai Printo Pack Pvt Ltd

Death closes all: but something ere the end,
Some work of noble note, may yet be done.
... Come, my friends,
'Tis not too late to seek a newer world.
Push off, and sitting well in order smite
The sounding furrows; for my purpose holds
To sail beyond the sunset, and the baths
Of all the western stars, until I die.
... Tho' much is taken, much abides; and tho'
We are not now that strength which in old days
Moved earth and heaven, that which we are, we are;
One equal temper of heroic hearts,
Made weak by time and fate, but strong in will
To strive, to seek, to find, and not to yield.

– *Ulysses*
Alfred Tennyson

CONTENTS

Foreword	ix
Prologue	xi
Early Years	1
History in the Making	24
On a Roll	62
A Billion-Litre Idea	94
Operation Flood	116
Tough Times	140
Step by Step	164
From Organisation to Institution	186
Life of Service	197
A Look Back	210
Postscript	232
Annexure	237
Index	243
Acknowledgements	250

the coterie's own interests and convert the institute into a profit-making private-management institution, like many others in the country, which cater to merely meeting the demands of private sector, particularly the corporates.

The people behind these attempts were people with considerable resources at their disposal. However, with appropriate and timely actions and with the help of those committed to IRMA's cause, I could keep IRMA from being derailed from its core mission – that of producing managers for India's under-managed rural organisations which believe in people-centred, equitable and sustainable development.

People may call me cold-blooded but I am very firm in setting correct precedents. This is a practice I will follow until the very end. As in life, so too in death. And because some things cannot be left to the people one leaves behind, I have already asked, for instance, that my body be cremated here in Anand. No special place and no special functions for me. As the poet, Alfred Tennyson, so sensibly said:

Sunset and evening star
And one clear call for me!
And may there be no moaning at the bar
When I put out to sea.

As I write these last few lines of my memoirs, my faith in the farmers of our country remains unshaken. Although I do feel betrayed by some whom I trusted, I continue to get support from unexpected quarters. The journey I began in Anand in 1949 still continues. I believe it will continue until we succeed.... Until India's farmers succeed.

31 December 2004

FOREWORD

INDIA HAS BEEN IMMENSELY FORTUNATE TO HAVE IN ITS MIDST A SELFLESS, committed and self-effacing nationalist like Dr Verghese Kurien, who has been able to dedicate his working life to improving the quality of life and health of millions of citizens of the poorest segment of India's population and transforming them into productive members of society.

This memoir allows us a fascinating view of the challenges, achievements and frustrations of Dr Kurien's full and colourful life. Dr Kurien was never one to mince words. 'You get what you see', and you know where you stand with him. It is so in his memoirs too. He has recounted instances as they were and he has not cloaked his feelings with niceties relating to certain people; at the same time he has been generous with his praise for others where he felt it necessary. In doing so, his memoirs reflect a true sense of events as Dr Kurien saw them throughout his life.

All of us should be proud of the achievements of Dr Kurien. A true visionary, he built a series of institutions which made India the world's largest milk producer, developed a logistic chain to produce and deliver hygienic and nutritious milk to millions and created the world's largest food marketing business and the country's largest food brand (Amul). He enabled India to nearly double its per capita milk availability and made India's dairy industry the largest rural employment provider. The cooperatives he created have also become powerful agents of social change in empowering women and in embedding democracy at the grassroots level in the country.

His memoirs reveal the experiences and inner feelings of a great nationalist who has made an enormous contribution to the development of rural India and who will leave his mark long into the future. The book should provide inspiration to many: that they should do something for their country. Dr Kurien's involvement with dairy engineering was a twist of fate. He would normally have pursued a career in science or engineering. What he, therefore, did for the dairy industry in India is truly amazing. One cannot help but wonder what India would be today if we had a thousand Dr Kuriens with this type of vision and with similar commitment, dedication and national spirit.

1 September 2004

Ratan N. Tata

PROLOGUE

TO MY GRANDSON

Anand, 2005

My dear Siddharth,

When did I write to you last? I have trouble even remembering! In today's fast-paced world we have become so addicted to instant communication that we prefer to use a telephone. But speaking on the telephone only gives us an immediate but fleeting joy. Writing is different. Writing – even if it is a letter – not only conveys our present concerns and views of the events taking place around us but it becomes a possession that can be treasured and re-read over the years, with great, abiding pleasure.

What is contained in the chapters that follow is, of course, more than a letter. You may not wish to read it all right away but, perhaps, a couple of decades or more from now, you will pick up these jottings of mine again and they will give you a deeper understanding of what I have done, and the reasons I pursued a life of service to our nation's farmers. You will then discover in them a valuable reminder of the days just before the world entered the twenty-first century. And you may want to share my memories with those of your generation, or even younger, to provide them a glimpse of the world your grandparents lived in and knew.

I started my working life soon after our country became independent. The noblest task in those days was to contribute in whatever way we could towards building an India of our dreams – a nation where our people would not only hold their heads high in freedom but would be free from hunger and poverty. A nation where our people could live with equal respect and love for one another. A nation that would eventually be counted among the foremost nations of the world. It was then that I realised, in all humility, that choosing to lead one kind of life means putting aside the desire to pursue other options. This transformation took place within me fifty years ago, when I agreed to work for a small cooperative of dairy farmers who were trying to gain control over their lives.

To be quite honest, service to our nation's farmers was not the career I had envisioned for myself. But somehow, a series of events swept me along and put me in a certain place at a certain time when I had to choose between one option or another. I was faced with a choice that would transform my life. I could have pursued a career in metallurgy and perhaps become the chief executive of a large company. Or, I could have opted for a commission in the Indian Army and maybe retired as a general. Or, I could have left for the US and gone on to become a highly successful NRI. Yet I chose none of these because somewhere, deep down, I knew I could make a more meaningful contribution by working here in Anand, Gujarat.

Your grandmother too made an important choice. She knew, in those early days, life in Anand could not offer even the simple comforts that we take for granted today. However, she ardently supported my choice to live and work in Anand. That choice of your grandmother to stand by me has given me an everlasting strength, always ensuring that I shouldered my responsibilities with poise.

Whenever I have received any recognition for my contributions towards the progress of our country, I have always emphasised that it is a recognition of the achievements of many people with whom I had the privilege to be associated with. I would like to stress even more strongly that my contributions have been possible only because I have consistently adhered to certain core values. Values that I inherited from my parents and other family elders; values that I saw in my mentor and supporter here in Anand – Tribhuvandas Patel. I have often spoken of integrity as the most important of these values, realising that integrity – and personal integrity, at that – is being honest to yourself. If you are always honest to yourself, it does not take much effort in always being honest with others.

I have also learnt what I am sure you, too, will find out some day. Life is a privilege and to waste it would be wrong. In living this privilege called 'Life', you must accept responsibility for yourself, always use your talents to the best of your ability and contribute somehow to the common good. That common good will present itself to you in many forms every day. If you just look around you, you will find there is a lot waiting to be done: your friend may need some help, your teacher could be looking for a volunteer, or the community you live in will need you to make a contribution. I hope that you, too, will discover, as I did, that failure is not about not succeeding. Rather, it is about not putting in your best effort and not contributing, however modestly, to the common good.

In life you, too, will discover, as I did, that anything can go wrong at any time and mostly does. Yet, there is little correlation between the circumstances of people's lives and how happy they are. Most of us compare ourselves with someone we think is happier – a relative, an acquaintance, or often, someone we barely know. But when we start looking closely we realise that what we saw were only images of perfection. And that will help us understand and cherish what we have, rather than what we don't have.

Do you remember when you accompanied me to the magnificent ceremony in Delhi in which our President awarded me the Padma Vibhushan in 1999? With great pride, you slipped the medal around your neck, looked at it in awe and asked me very innocently if you could keep it. Do you remember the answer your grandmother and I gave you? We told you that of course, this medal was yours as much as it was mine but that you should not be satisfied in merely keeping my awards – the challenge before you was to earn your own rewards for the work that you did in your lifetime.

And in the end, if we are brave enough to love, strong enough to rejoice in another's happiness and wise enough to know that there is enough to go around for all, then we would have lived our lives to the fullest.

I would like to dedicate these musings to you, Siddharth, and to the millions of other children of your generation in our country, in the hope that upon reading them you will be inspired enough to go bravely out into your world and work tirelessly in your chosen field for the larger good of the country, for the larger good of humanity. Remember, the rewards that come to you then are the only true rewards for a life well-spent.

With my fondest love,

Yours affectionately,
Dada

EARLY YEARS

'DR KURIEN, ONE FINAL QUESTION,' ASKED A YOUNG AND EAGER JOURNALIST not so long ago: 'What are your plans for the future?'

I remember being somewhat amused by this staple media question. There comes a time in a person's life when the future starts to become a little irrelevant; each new day is like a bonus. My response, I think, was appropriate; even today, it befits my eighty-three years. 'At my age,' I explained, 'one does not really have a future. One only has a past.'

Looking back, I realise that I have been one of the lucky few to have lived a life so busy, so packed with plans and purpose that eight decades seem to have flown by in a trice. I think I can take pride in the fact that my mission is, by and large, accomplished. The time has come for me to step down and place the reins of the many offices I have held in firmer hands. I, now, take pleasure and pride in handing over the baton to young and able successors.

Over the years it has been my privilege to hold a variety of offices. From April 1950 to July 1973, I was the Manager and then the General Manager of the Kaira District Cooperative Milk Producers Union Ltd (popularly known as 'Amul'). From October 1973 until October 1983, when I reached the age of superannuation, I was the Founder Chairman-cum-Managing Director of the Gujarat Cooperative Milk Marketing Federation (GCMMF). Since then, I have continued as the elected Chairman of

GCMMF. For thirty-three eventful years between 1965 and 1998, I served as the Chairman of the National Dairy Development Board (NDDB), an organisation promoted by the Government of India. Many different governments came to power in these three decades and each allowed me to continue, despite my reputation of often crossing swords with bureaucrats and ministers. As I write these memoirs I continue as the Chairman of the Institute of Rural Management (IRMA), Chairman of the GCMMF and Chairman of the National Cooperative Dairy Federation of India Ltd (NCDFI). These positions allow me to continue to serve the interests of the nation's dairy farmers and rural people.

Often, these days, while sitting at my favourite desk (which my colleagues so thoughtfully packed off with me when I retired from NDDB) in my tastefully decorated office, I glance out of the wide glass windows with pride at the meticulously kept, luxuriant sixty-acre IRMA campus. It is an expanse of verdant, undulating lawns. I see faculty members walking to their lectures, students engrossed in conversation.

The equally impressive grounds of the forty-acre NDDB complex lie barely a stone's throw away. A couple of kilometres from the complex are the splendid premises of Amul and close by, of GCMMF.

Off the Mumbai-Delhi National Highway, about forty kilometres from Baroda, at an inconspicuous turn, stands a modest blue-and-white board with an arrow pointing towards Anand, announcing: 'The Milk Capital of India'. It leads you to the offices of AMUL, NDDB, GCMMF, IRMA and to the dairy farmers they serve and represent, which have earned this little town such a lofty label. These institutions make Anand the pride of Gujarat's milk producers and of the entire nation. But it was not always like this. There was a time, approximately fifty years ago, when this was just another dusty, sleepy little town like hundreds of others that dot our countryside. I have often said that it was a sheer quirk of fate that brought me to this small town in Gujarat to which my life has become so inextricably linked.

I was born on 26 November 1921, in Calicut, Kerala, and was the third of four siblings. I was named 'Verghese' after my uncle, Rao Saheb P. K. Verghese, who had made a notable contribution to public life in his home town, Ernakulam. My father, Puthenparakkal Kurien, served as a civil surgeon in British Cochin. My mother was talented – she played the piano exceptionally well – and highly educated. She came from an illustrious family which laid great store by learning. In fact, by and large, the Syrian Christian community to which I belong, gaining much from the British policy of educating Indians, had achieved a high level of literacy.

When I turned fourteen, I joined Loyola College in Madras (now known as Chennai), to study science. I was very young for my class but I learnt to cope with the studies. When I completed college, in 1940, I was still too young to get admission in an engineering college, so I did an extra degree, a B.Sc. in Physics at Loyola. After this I enrolled at the Guindy College of Engineering, also in Madras, which then served the entire South India. I was very young when I joined college and had to manage more or less on my own. I learnt to fend for myself and became independent very early in life.

I enjoyed my years at both the colleges, for not only was I academically inclined but I revelled in sports too. My father had been an athlete of repute in his youth, earning the nickname 'hundred-yard dash Kurien' and he passed on his love of sports to me. I represented my college in tennis, badminton, cricket and boxing, without in any way harming my excellent academic record. Boxing, I recall, was serious business in college and took up a lot of my time. We trained rigorously with the coach and would often emerge full of cuts and bruises, black eyes and swollen lips, and unable to eat. I also joined the University Training Corps

(UTC) and I am still proud of being selected as the outstanding cadet of the 5th battalion of the Madras UTC. Our Adjutant was Captain K.S. Thimmaiah, who later rose to become a distinguished General. My years at college were busy beyond imagination.

While I underwent training with the UTC, I became quite enamoured with the discipline and exactitude inculcated by army life, and began toying with the idea of joining the army. Unfortunately, my father died prematurely, when I was only twenty-two. But among our small Syrian Christian community, family ties are extremely strong and I was never allowed to suffer for too long the pain and sense of terrible loss that accompanies the death of a parent. On hearing of my father's death, my maternal grand-uncle, Cherian Matthai, came as soon as he could and took my mother and all of us to Trichur (now Thrissur) where he lived in a large, well-appointed home.

Cherian Matthai was the Director of Public Instruction of the Cochin state and, in the community, was fondly referred to as 'Matthai Master'. He was the grand old man, the patriarch of our family and the eldest brother of John Matthai (who later became India's Finance Minister). Since John Matthai was of my mother's age, they had been brought up like siblings in the same family home in Calicut. 'Matthai Master' never married but he looked after the entire family. He sent his brothers and sister to England for education. He lived in a sprawling house on a hundred-acre estate, with a boat club, gymnasium, golf course and excellent cooks. Ever since I can remember, even before my father's death, we used to spend our summer vacations with 'Matthai Master' and what memorable days those were! Thanks to this large, closely-knit family, I have intensely happy and glorious memories of childhood.

After I finished my engineering my mother, who had managed to distract my mind away from the army, now deftly guided me towards trying my luck with the Tata Iron and Steel Company

(TISCO). I relented and, in 1944, I was selected by TISCO as a graduate apprentice. This was considered a very prestigious selection then because the company took only ten 'A class' apprentices. I was unaware at that time that very soon I would be confronted by a peculiar problem.

In those days Sir Jehangir J. Ghandy was the Managing Director of TISCO. He reigned as the uncrowned 'King of Jamshedpur'. Such were his powers that if he smiled at you, you were made for life; if he frowned, you were doomed. Unfortunately for me, my uncle John Matthai, as a Director at Tata Industries, was his boss and had requested him to consider my application for apprenticeship 'if found competent'. Sir Jehangir had no choice but to obey, although I do believe I could have got in on my own merit. I was posted to Jamshedpur and even if I say so myself, I was a competent engineer and a good apprentice.

As if John Matthai putting in a good word for me was not bad enough, one day he committed the unforgivable mistake of visiting me at the apprentices' hostel. We apprentices were the lowest in the officer category and no senior officer ever visited our hostel. Everybody noticed when John Matthai came to see me. They realised who I was. They were convinced that I would, inevitably, be marked for rapid promotions in the Tata Group. Suddenly everyone was extremely good to me and very careful around me. I found this unbearably oppressive and knew that I would have to do something about it soon.

On my uncle's next visit to Jamshedpur, I told him politely, 'I don't want to stay here. I want to get out. I am no longer Kurien. Now I'm merely the boss's grand-nephew.'

'Very, very commendable,' he said, nodding his head. 'But also extremely stupid. I'm told you are the best apprentice. You will certainly go right to the top here.'

I was adamant and informed him that I had already applied for a scholarship from the British government for higher studies.

He was extremely unhappy with my decision and tried very hard to persuade me to stay on with the Tata Group where, no doubt, my career would flourish. But I had made up my mind and that mind told me in no uncertain terms that I must get out of this situation.

The British government had announced a scheme to select about five-hundred young Indians to send abroad for specialised training to England, New Zealand, Australia, Canada and the US. I applied, hoping to go abroad and get a Master's degree in metallurgy and nuclear physics. I was one of the lucky ones to get a call for an interview with the government's scholarship selection committee.

During the interview on the specified date, the Chairman of the selection committee, after inviting me to sit down, asked me only one question: 'What is pasteurisation?'

I did not know exactly and I replied hesitantly but quite honestly, 'I don't know the process but I think it has something to do with sterilising milk ... .'

'Correct,' he said. 'You are selected for a scholarship in dairy engineering.'

I was taken aback. 'Dairy engineering?' I asked incredulously. 'Can't you give me metallurgy or nuclear physics?'

'No. It's either this or nothing. Make up your mind,' he said.

I was in a bit of a quandary, but I knew I had to find an exit from Tata Steel. I accepted the scholarship to go to the US and qualify as a dairy engineer for the Government of India's Ministry of Agriculture.

Since I had unwittingly revealed that I did not know anything about dairying, before I left for the US, I was sent for eight months to what was then called the Imperial Dairy Research Institute in Bangalore (later the National Dairy Research Institute of India) for a formal introduction to milch cattle and to try and understand the fundamentals of dairying. As soon as I reached the institute in

Bangalore I knew I had made a serious mistake in leaving the Tatas. The institute's officers did not take too kindly to me. I was an outsider; I knew nothing about dairying and yet I had been selected for a coveted scholarship. Nobody bothered to teach me anything and one of the instructors in particular – Kodandapani – took a special dislike to me. But I had burnt my bridge and there was no turning back. I struck up a friendship with two dairy technologists, A.T. Dudani and Pheroze Medora, and spent a considerable time at restaurants, movie halls and generally having a good time. As far as I was concerned, I was merely marking time till I got my scholarship to go abroad.

In the winter of 1946, I left for the US abroad a 'Liberty' ship to join Michigan State University, ostensibly to study dairy engineering. What I actually studied there was metallurgy and nuclear physics. To satisfy the Government of India, I took some token courses in dairy engineering. The first atom bomb had been exploded and I saw nuclear physics as an area with tremendous scope. Dairying did not figure anywhere on my horizon.

Those days, Michigan State University was considered the world's best place for dairy engineering. As luck would have it, Pheroze Medora, my friend from the Imperial Dairy Research Institute in Bangalore, also joined the university, as did his friend, Harichand M. Dalaya, who was later to become my close and valued colleague. Medora and Dalaya had studied together at the Agriculture College in Poona (now Pune). Dalaya had come to Michigan State University on his own steam since he came from a well-to-do family. Two other Indian students – Hussain and Mansoor – also became close friends. We were a group of five Indians – Medora was a Parsi, Hussain and Mansoor were Muslims, Dalaya was a Hindu and I was a Christian. A veritable object lesson in national integration.

My easy-going fun-filled lifestyle worried Dalaya tremendously and he was convinced that I would end up in a mess. As far as he was concerned, there was no hope for me. He took it upon himself to advise me, to berate me when he thought I was spending too many evenings out, not studying enough and not spending the mandatory hours in the laboratories. Finally, one day when I could not take his nagging any more, I told him, 'Dalaya, you get your degree, I'll get mine and then we'll see. You'll get it huffing and puffing, I'll get mine laughing.' And I proved this to Dalaya by getting my Master's degree with distinction even while I enjoyed life to the fullest.

Those were good, productive days at the university. I pursued tennis and even won a championship. I spent stimulating evenings with my friends, debated and argued vociferously; whenever the competence of the Third World was questioned or a racist remark was made – not at all unusual in those days – I put the 'natives' soundly in their place.

Of course, I studied too. My research was on a fascinating subject – heredity in cast iron – and I was totally engrossed in it. There had long been a belief among foundry men that there is 'heredity' in cast iron. My research examined whether this was mere fiction or a fact and my thesis proved that it was indeed a fact – that heredity did exist. In a way it showed how you cannot get away from the past – even if it is in cast iron – so when you melt down cast iron and make it into something else, its past exists even in the face of apparent change.

During my research, my professor and I made what is known as colloidal iron, where the carbon is round in shape and not in flakes. This meant that cast iron, like steel, would also have the ability to stretch. These findings were very exciting and would have been path breaking. However, one day my professor informed me that somebody else had beaten us to it. We had come so close. We could have become millionaires! In a way, it was good because if I had become a millionaire, I would not have left the US.

Along with my four friends, I returned to India in 1948, after receiving a Master's degree in metallurgy and nuclear physics. It was the end of a wonderful, carefree chapter in my life. When the five of us returned home, we saw with horror how the British had divided our country and it had been divided without even seeking our permission. Hussain and Mansoor moved to the newly created Pakistan, and the rest of us got down to the business of getting on with our lives in India. I duly reported to the government, which had sent me on the scholarship, and was instructed to get in touch with the Ministry of Education in Delhi. By this time, my uncle John Matthai, who used to be a Director at the Tata Industries, had become the Finance Minister of independent India. I stayed with him in Delhi.

Those days, as a cocky, foreign-educated young Indian, I dressed rather nattily. My attire might have offended the sartorial sensibilities of some: my favourite clothes were a green shirt, yellow pants and a green felt hat. Thus decked out, I went one morning to see the Under Secretary, Education. The Under Secretary looked me up and down and said, 'Oh, so you are Kurien? You are one of the lucky ones. Most of the others have no jobs but for you we already have a job lined up. You will have to report to a place called Anand.'

'Where is this Anand?' I asked.

'It's somewhere near Bombay,' explained the Under Secretary.

At Michigan State University, the Dean of Engineering had taken a liking to me and when I finished my studies he recommended me to a company called Union Carbide. Union Carbide had already offered me a job with a basic salary of Rs 1,000 in Calcutta, where they were setting up a factory. Surely, then, I did not need this inconsequential government job? The Under Secretary was not very pleased with my demeanour, possibly even less with my sense of fashion. So when I informed him that I was not really interested in taking up the Anand assignment, he got rather agitated.

'How can you talk like this?' he exclaimed. 'I will sue you for the Rs 30,000 that we spent on your higher education if you refuse to take up this job at Anand.'

It seemed that I had no option. There was no way I could raise the money to repay the government. I would have to go to Anand. As I was about to leave the office, the Under Secretary asked me to wait while he gave me my appointment letter. I had promised my aunt that I would be home for lunch and I was already running late, so I told him that I could not wait. He grew even more furious. 'You cannot wait for your appointment letter because you have a lunch appointment? Young man,' he scolded, 'you will not go too far in life, there's no doubt about that. Anyway, give me your address in Delhi so that I can have the letter delivered to you.'

I wrote down my address, which was 'C/o the Honourable Dr John Matthai', and sped home for lunch.

'So,' asked my uncle with a smile, 'did you get your release?'

Seeing him in a relatively good mood, I asked him if he would help me get the release. He refused emphatically, taking great pleasure in pointing out, 'I told you not to leave the Tatas. I told you not to take the government scholarship. You rejected my advice. You wanted to build your future with your own efforts. So go build your future. I will not help you. You have made your bed, my boy. Now go lie in it.'

I felt cheated by the Government of India when I learnt later that I had actually been selected for a job as a dairy engineer at the Imperial Dairy Research Institute in Bangalore. This was a Senior Class I position. But when I returned to India, the officers at the dairy department got together and decided to post me to Anand and send a diploma holder, Kodandapani, back to Bangalore in my place. For a long time I bore a grudge because I felt that I had been cheated out of the research institute job which should have been mine. Looking back, of course, I do not know what would have been in store for me if I had been given what was rightfully my due. Possibly the best that could have happened was that I would have ended up as Director of the National Dairy Research Institute and retired. Instead, I was sent to this strange place called Anand, where, unknown to me, a far more challenging life lay ahead. The Dairy Development Adviser to the Government of India, Zal R. Kothavala, later told me that in my confidential report, my superiors had written: 'Unlikely to be an efficient officer.'

Meanwhile, when the Under Secretary realised that I was John Matthai's nephew he recommended a higher salary of Rs 600 a month for me. This required the finance ministry's approval, which the Under Secretary felt would certainly not be difficult. Little did he know my uncle.

'The finance ministry will never agree,' declared my uncle. My aunt asked him why it bothered him if I was going to get a better salary. 'It is my job to see he does not get it,' replied John Matthai. Such were the principles and standards of the Government of India those days.

Quite unknown to me, a unique and revolutionary experiment had been unfolding for some time in the little town of Anand where I was soon to make my home. It was here that a fledgling milk cooperative was desperately trying to hold its own. At that time, the milk business in the region was almost totally controlled by a shrewd and remarkable Parsi gentleman called Pestonjee Edulji. He had never been to school, had no education, but was clearly a canny entrepreneur. The manner in which Pestonjee came to monopolise the milk business in Anand is an interesting and instructive story.

Around 1942-43, a large number of Britishers stationed in Bombay (now Mumbai) fell sick. After an intensive investigation, the authorities identified the root of the problem as the milk they were drinking. A sample of this milk was sent for testing to a laboratory in London since the British would not trust any Indian laboratory. There was a one-line response from London. It said quite simply: 'The milk of Bombay is more polluted than the gutter water of London.' Thus, when milk in Bombay was found unfit for consumption, the British government felt compelled to create the post of a milk commissioner. A milk department was established to work out a scheme to improve the quality of milk coming into Bombay.

Interestingly, this was the first time the government had intruded into our country's milk business. It is even more interesting that the British government was forced to do so not because of any socialist ideals (which independent India subsequently adopted) but because the private sector had messed things up so much that it became necessary for the government to step in.

The first thing they did was to look for a nearby source of adequate milk supply and it was inevitable that they stumbled upon Kaira district, which was famous even during those days for its dairy industry. Around 1895, an Englishman had started a butter factory some fifteen kilometres north of Anand in this district. A German also set up a casein factory in the vicinity and in 1926, Pestonjee Edulji put up a large factory manufacturing butter and marketed it rather cleverly under the Western sounding brand name of 'Polson'. Polson butter soon became a household name.

A New Zealander – Foster – was the Manager at Polson. These pioneers – the Englishman, the German and the Indian – were creating a market for the milk from Kaira district. As always happens when a market is created, the producers react. In this case, the farmers of Kaira district responded and milk production increased. In Kaira district, therefore, before there was even any mention of a milk cooperative, the farmers had increased their production to such an extent that the district had already become the largest and best-known milk pocket in the entire region.

Zeroing in on Kaira, the British government of the Bombay state then asked the owner of Polson dairy whether it was possible for him to send milk from Anand to Bombay city – some 350 km away. Never before had liquid milk travelled such long distances in a hot country like ours. But Pestonjee was not one to throw up his hands in despair. He experimented. He pasteurised milk in his cream pasteuriser and transported it to Bombay in a rather primitive fashion – in milk cans wrapped in gunny bags with chilled water poured on the cans. He found that it reached Bombay in fairly good condition. This was the beginning of the government's Bombay Milk Scheme – probably the first milk scheme to supply liquid milk to any distant city in India.

Bombay thus became a market not just for milk products but also for liquid milk. It was an additional market for Kaira district and an incentive for its dairy farmers. Bombay – the commercial and industrial capital of India, the city of the rich and famous – was now directly linked to Kaira. It was certainly an enormous stimulus, which further increased the production of milk in Kaira district. These were all undeniably vital pioneering efforts in the district's milk industry.

Pestonjee – or Polson – found this arrangement with the British government quite satisfactory. He demanded a processing charge from the government, which the government agreed to give. He demanded additional equipment as a grant and that, too, was provided. Then Polson made one more demand. He told the government that for him to do his job more efficiently and supply Bombay with more milk, they should pass a legislation whereby in all the villages around Anand no one besides him could collect milk. This, too, was done, which meant that contractors appointed by Polson monopolised milk procurement.

Polson was happy, the Milk Commissioner of Bombay was satisfied and, above all, the milk contractors and merchants were ecstatic. The dairy farmers, on the other hand, were dejected and miserable. The farmers very soon realised that the increased price, which Polson obtained from the government, went into the pockets of Polson dairy and worse, much of it went into the grasping hands of the milk contractors. Only a miniscule amount of the increase in price reached them.

This was the time when India was struggling for its freedom from British rule. Kaira's farmers complained about their exploitation to Sardar Vallabhbhai Patel, a prominent leader of the freedom movement and Deputy Prime Minister of independent India, who came from Karamsad village just a few kilometres from Anand. Sardar Patel was the man who abolished the rule of the rajas and maharajas overnight and welded the country into one nation; he was the 'Iron Man of India', a great administrator and patriot. He firmly believed that a revolution in marketing the farmers' produce – which would be beneficial to the farmers – was necessary. Sardar Patel was convinced that in order to save themselves, the farmers needed to control the procuring, processing and marketing of milk.

Sardar Vallabhbhai Patel's vision has always been a source of great inspiration. After fighting for and winning freedom, he recognised that independence was more than a political task. He knew that our rural people could never become really free until they were liberated from the exploitation of moneylenders, from the social ills and burdens of caste and class. Sardar Patel believed that the way to address these problems was to build rural institutions, and institutions of research and teaching, that would serve the farmers' economic interests; institutions that would cater to the needs of the rural people.

Sardar Patel urged the dairy farmers to organise milk cooperatives, which would give them control over the resources they generated. He assigned Morarji Desai, his deputy, to coordinate this effort. At a meeting of the dairy farmers, Morarjibhai asked for volunteers to serve as chairman of the organisation. A few people volunteered, but Morarjibhai looked around and spotted Tribhuvandas Patel sitting quietly in the gathering. Tribhuvandas was then a young and committed freedom fighter and the elected Vice-President of the Kaira District Congress Committee.

'Don't you want to be the chairman?' Morarjibhai asked him.

'No, Saheb,' replied Tribhuvandas. 'I've just come out of jail for the fourth time and my health is not so good. I just want to go home and recover. Besides, I don't know anything about the dairy business.'

Morarjibhai heard him out and said, 'So you don't want to be the chairman? In that case, you shall be the chairman.' And Tribhuvandas was selected as the Chairman of the Kaira Cooperatives. Morarjibhai probably believed that if somebody wanted to be the chairman badly enough, then he would definitely have some vested interest, and that was certainly not right.

Tribhuvandas launched the exercise of organising the dairy farmers and he soon managed to form a couple of cooperative societies. While the farmers were willing to take their leader's advice, taking on the vested interests was not easy. Apart from anything else, since milk is such a highly perishable commodity, the farmers of Kaira were vulnerable, left with little choice but to accept whatever the contractor paid. Exploitation increased. The cooperatives supplied milk to Polson who would process it and supply it to Bombay. Predictably, problems cropped up. Polson would 'discover' flies in the milk; he would say it did not smell fresh; he would not give a fair fat test or payments. He tried every trick he could to break the cooperatives. What was the next step then?

It was then that the farmers and their leader, Tribhuvandas Patel, again went to meet Sardar Patel, who gave them only one bit of pithy advice. 'Polson ne kaadhi mukho, remove Polson,' he said. It sounded a simple enough directive but it was not so easy to accomplish. He also told them that if they wanted maximum benefit from the Bombay market, they had no other option but to cooperatise the dairy movement in their district and, above all, to own their dairy. He gave them this advice in 1945, before India was free. He warned them that the British would not like it because they would see in this a ruse to embarrass their government; they would think that he, Sardar Patel, was organising the farmers against the British; that by forming the cooperative, he had political ends in mind — which he did not. Knowing fully well how difficult it would be, Sardar Patel cautioned the farmers of Kaira district that if they wanted to achieve all this they must be prepared to fight, not forgetting that in any fight there are losses. 'The losses will be yours, not mine,' he told them. 'But if you are prepared to struggle, to bear the losses, to fight the Milk Commissioner of Bombay and his department, then I am prepared to lead you.'

As the farmers were fed up of being exploited they readily agreed. What is of great significance is that the cooperative dairy movement in Kaira district started as a farmers' initiative against the government, the Milk Commissioner and the department of Dairying of the British Government of Bombay.

Once the farmers decided that they would fight the government, Sardar Patel sent Morarji Desai to carry the struggle forward. In January 1946 Morarji Desai, who was then the Secretary of the Gujarat Provincial Congress Committee, came to Kaira district and held the first meeting under a banyan tree in Chaklashi village, some ten kilometres from Anand. At the meeting only two resolutions were passed. The first said that no milk would be sold to Polson dairy and the second said: 'We propose to have a cooperative in each of our villages in Kaira district and a union of these village cooperatives at Anand will handle the processing of milk. This will enable the farmers to gain control over the procurement, processing and marketing of our milk.'

It was Morarji Desai who led this initial farmers' movement against the Bombay Milk Scheme (BMS). He demanded that the BMS accept milk from the farmers' cooperative directly and not merely the milk supplied to it by Polson dairy. As was expected, the demand was rejected and so he declared that the farmers would go on strike against the government's Bombay Milk Scheme. This was the famous fifteen-day milk strike of Kaira district, during which all the milk that was collected by the farmers was poured on the streets but not a drop was given to Polson. Polson's milk collection came to a grinding halt and the BMS collapsed.

At this stage the Milk Commissioner – an Englishman – and his deputy, a famous dairyman of India, Dara Khurody (who later built Bombay's Aarey Milk Colony), decided to visit Anand. They saw that the farmers were adamant, their spirits were very strong and their strike was unlikely to end. Khurody advised the Milk Commissioner to concede to their demands. 'Look at their leader Tribhuvandas Patel,' he said. 'He wears a Gandhi topi, he cannot speak English. How is he going to handle this milk business? This is not New Zealand or Denmark. This is India. Milk business is a technical thing. Do you really think the cooperatives in Anand can succeed? Concede to their demands; they are only doomed to failure.'

This was exactly what Tribhuvandas Patel was hoping for. Throughout that year, he trudged tirelessly, mile after mile, from village to village, almost single-handedly persuading the farmers of Kaira district to form cooperative milk societies. Towards the end of that year five societies were registered and by December 1946 Tribhuvandas and the dairy farmers registered the Kaira District Cooperative Milk Producers Union Limited (KDCMPUL).

Fortunately for us, we won independence soon after and Sardar Patel became the Deputy Prime Minister of a free India.

Tribhuvandas Patel went to Delhi to meet Sardar Patel to inform him of the persisting difficult situation in Anand. He also mentioned that the farmers felt an acute need to have their own dairy. There already existed in Anand a very old government research creamery, which had been built in 1914 and had fallen into disuse. This was the same dairy which had once been used to make cheese for the British troops in Mesopotamia (near Iraq) during World War I and which the National Dairy Research Institute (NDRI) had acquired from the Government of India on lease. It may be apocryphal, but the story went that the cheese that was made here and consumed by the British troops killed more British soldiers than did the enemy!

Sardar Patel told Tribhuvandas to meet the then Minister for Agriculture – Rajendra Prasad – and ask him to give the dairy to the farmers' cooperative. However, his bureaucracy opposed it, saying that government property could not be simply given to anyone gratis. Therefore, on the sage advice of bureaucrats, a part of this dairy along with some of the machines, was given to the Kaira Cooperative Union on a rent of Rs 9,000 per year. The rest of the dairy remained with the NDRI, which used it for its own research.

This was a World War I vintage dairy and, understandably, in an awful state. It was a curious assortment of ancient machines: a boiler, which, when the pressure was built up would start a steam engine which would then drive a shaft across the length of the dairy, painfully working the ancient mechanism. There were all kinds of pulleys and belts and pumps. All very impractical, antiquated stuff.

It was around this stage in the life of the Kaira Cooperative Union that I happened to arrive at Anand. As I have often said later, I was thrown into this little dusty town in Gujarat and into the lives of the dairy farmers of Kaira district by what I consider to be a sheer accident of fate – what turned out to be a strange pre-planned act of destiny. I had always imagined that I was cut out for 'bigger and better things' – for a glamorous, fast-paced life in a big city, a job with a prestigious firm and the pleasures of the luxurious lifestyle that go with it. Anand did not figure anywhere in my scheme of things. But I had to honour the contract with the Government of India which had enabled my higher studies in the US and therefore, here I was in Anand, a fish out of water.

~

It was Friday, 13 May 1949, when the train from Bombay pulled into the ramshackle station at Anand. There to receive me was none other than Kodandapani, looking quite pleased with himself since he could now return to Bangalore. Accompanying him was the elderly, friendly caretaker of the government research creamery, known to all as 'Barot Kaka'. They escorted me to my new office at the creamery, which was then being run by the NDRI. Kodandapani very considerately warned me not to take charge on that day as Friday the 13th is considered an unlucky day. He advised me to wait for a day. 'No,' I said. 'I don't like what I see here. Let me take charge today and allow things to go wrong. I'm not interested in staying here too long.'

In 1949, Anand had a population of approximately ten thousand people. It was the antithesis of my life in New York, which I had then just left behind. New York was bustling and busy, Anand was dull and sleepy; New York was vibrant and liberal, Anand was excruciatingly unexciting and conservative. It was so conservative that nobody was prepared to rent me a house or even a room. I seemed to have all the possible disqualifications. I was an outsider, a Malayali. In addition, I was a Christian and a non-vegetarian – an outrage for the strictly vegetarian Gujarati community. And to top it all – I was a bachelor! Which self-respecting Gujarati family would let out a room to an unmarried Malayali Christian?

After days of searching for a room, I managed to rent an abandoned garage of the house next to the dairy, which was occupied by the research institute's Superintendent. What a twist of fate – from comfortable lodgings in glittering New York to a garage in dreary Anand. The garage had a large, greasy pit in the middle, where mechanics must have once stood to tinker with the undersides of cars. But I was an engineer and I could not allow such things to get in the way of basic comfort. I filled up the pit. There were no windows, so I created them. There was no bathroom so I put up three corrugated sheets and a makeshift bathroom was ready. This is how I started my life at Anand. Nobody gave me anything – and in the long run, that did me a lot of good.

I hated Anand. I wanted to run away but I could not because the Government of India had paid for my education in the US and I was under contract to work wherever they sent me, for a period of five years. I did not have the money to pay the government to get a release from the bond. It was just my misfortune that they sent me to Anand. But soon life began to take on a pattern of sorts. I would report to work every morning, tinker around with the machinery, try to find something to do in a place where everyone seemed to believe that you could get away without doing any work. I started smoking heavily, did not bother to shave most days and as a result looked shabby and unkempt. My felt hat and colourful clothes had no use here and I took to hanging around in my working clothes – khaki overalls. Most evenings were spent playing cards with Bala Rao, the Chief Chemist at the dairy, and some young men from the nearby Beedi Tobacco Research Station.

The highlight of my life was to go back 'home' to my garage for dinner. My elder brother, who then worked with the Birlas and lived in Hyderabad, had taken pity on me and sent me his trained cook-cum-butler, Anthony. Every evening without fail, at dinnertime, Anthony appeared before me in my tiny garage like a genie from Aladdin's lamp. He would be dressed up in his impeccable white uniform, with a sash and a turban perfectly in place, fix me with an unwavering look, and announce in all seriousness: 'Master, dinner is served.' And, indeed, my small, rickety table would be flawlessly set in one corner of the derelict garage. It must have looked like a scene out of a farcical comedy but it was such touches that kept my sense of humour intact during those dreary days. Anthony – all credit to him – stuck it out with me for many years at Anand.

The government research creamery where I was supposed to work actually did not conduct any research whatsoever. Our assignment here was to manufacture small quantities of milk powder from buffalo milk. This hardly required any research but in their typical bureaucratic way, the government had managed to turn it into a production. In reality not even the first step of the project had been set in motion. When I came to Anand even the roller-dryers were not working due to a minor fault which took me exactly ten minutes to set right. It certainly did not need someone with a Master's degree in engineering to be able to do that. Once the roller-dryers were working, producing milk powder was hardly an achievement.

My supervisor was a lazy, good-for-nothing fellow. He would come to work at eleven o'clock in the morning, use the institute's staff as his personal servants and do no work. Every morning he would saunter into my office to have a cup of tea and pass the time. One day I suggested to him that since we were now producing milk powder and we already had five tons of it, we should think of selling it. He began to complain about how difficult that would be as nobody would want to buy it. I thought that since it was good quality milk powder, any biscuit manufacturer would be happy to buy it from us. My supervisor was extremely sceptical. Those days I would frequently escape to Bombay, stay at the Taj Hotel and live it up for a few days. This was my rest and relaxation time and it was sacrosanct because it was the only thing that kept me going in this godforsaken place. Nothing would persuade me to miss this reprieve. I offered to try and sell the milk powder during my trip to Bombay. My supervisor agreed and on my next jaunt to the city I approached a biscuit manufacturer, showed him the sample milk powder I was carrying and negotiated a price of Re 1 per pound. He gave me a purchase order for five tons that we had in stock and with that in my pocket, I returned to Anand.

It was as simple as that. A little bit of creative thinking and initiative on my part and the work was accomplished. It made the sheer incompetence of my colleagues at the research creamery in Anand even more intolerable to me. I could see that they had no interest in doing anything, not even the most elementary of jobs. They employed twenty people to run two small roller-dryers when in any other country twenty such roller-dryers were run by one man. I was the new dairy engineer to the Government of India Research Creamery and I realised very soon that I had no work at all. My frustration at this deadening job began rising and I started to write to the Ministry of Agriculture in Delhi every month, submitting my resignation, saying that I was drawing a salary of Rs 350 for doing no work and instead of wasting government money I should be allowed to go. After some eight months of this they must have felt that I was becoming a nuisance and they finally wrote back accepting my resignation.

During those eight months, however, I looked around and said to myself that since I found myself placed in this unhappy situation, I must find something to do. This is what anyone with a good education would do; otherwise the so-called good education is worthless.

At that time, a cooperative was a completely alien concept to me. But while I knew nothing about cooperatives, I was quite intrigued by the band of tenacious dairy farmers and their leader,

Tribhuvandas Patel, at the Kaira Cooperative Union next door to our research creamery. It seemed to me that they were struggling against an impossible situation and I secretly felt that they would get nowhere with all their struggles. Yet I could not help but admire their drive and commitment to the cause. In my free time – which I had in plenty – I began helping this unique group of people next door. This was my first encounter with Tribhuvandas and my first introduction to Kaira's milk cooperatives.

HISTORY IN THE MAKING

UNDER THE LEADERSHIP OF TRIBHUVANDAS PATEL, THE KAIRA DISTRICT Cooperative Milk Producers Union Limited (KDCMPUL) began operating from the dismal half of the government creamery they had rented. They laboured painfully to get the rusty World War I equipment to function. However, the machines would break down at regular intervals. Since I was a 'foreign-returned' engineer who worked at the government creamery next door, Tribhuvandas began asking me for 'expert' advice. In any case, I had hardly any work to fill my time at the dairy and instead of whiling away the hours playing cards with others as idle as myself, I would walk across to the cooperative and fix the machines for them. This became a pattern until one day after I had patched up one of the machines for the umpteenth time, totally fed up, I told Tribhuvandas: 'Scrap this dairy. With this antiquated equipment you will never make a success of the business. I can continue to help you to fix it up but this is no way to run your milk cooperative. Why don't you borrow some money and buy modern dairy-plant equipment?'

Tribhuvandas went to his colleague, Maganbhai D. Patel, the Vice Chairman of the cooperative, for advice. Earlier, Maganbhai had sent Prof. Pandya, Agriculture Engineer and Professor of Agriculture at the cooperative, to help. I spent an entire day explaining to Prof. Pandya what needed to be done with the machinery. He took extensive notes and towards the end of the day, he closed his papers saying, 'I can't do it. I am unable to understand this.' Now there was no one else to deal with it. Therefore, when Tribhuvandas sought Maganbhai's opinion, he said, 'You have no alternative. Follow Kurien's advice. Borrow some money.'

Tribhuvandas reverted to me for advice on the type of new machinery that was required. I suggested that they should get a plate pasteuriser, which would cost them around Rs 40,000. It was an enormous amount. Tribhuvandas managed to borrow the amount from his large-hearted, affluent brother-in-law, with a promise that the cooperative would repay every paisa.

His brother-in-law reassured him: 'Tribhuvandas, don't worry. You take the money. You and your mad ideas – there's no way this dairy will work. Anyway I don't need this money so you don't have to worry about it.' Tribhuvandas, a man of integrity, paid him back every paisa, with interest, once the cooperative was in a position to do so.

Tribhuvandas requested me to order the plate pasteuriser on behalf of the cooperative on my next visit to Bombay. After consulting Dalaya, I decided to order the machinery from Larsen & Toubro (L&T). With cash in my pocket, I walked into the office of the L&T Manager, Axil Petersen, and told him that I wanted to order a Silkeborg Pasteuriser. I must have been quite a sight – unshaven, shabby and unkempt. Petersen looked at me a little oddly and asked me if I knew how much the machine would cost. I assured him that I did. At which, with the typical arrogance of the coloniser, he asked to see the money. I pulled out a thick wad, of Rs 40,000, from my pocket and rather dramatically threw it on his desk: 'You can see it right now,' I added.

Petersen was obliged to take my order and promised to deliver the plate pasteuriser as soon as possible. With the job completed, I enjoyed the rest of my holiday before returning to the boredom of my job.

Every month I sent my resignation letter to the Secretary, Ministry of Agriculture, Government of India, each time pointing out that the government was paying me a salary for no work. Around this time my friend Medora, now a chemist with the Bombay Milk Scheme's laboratory at Anand, asked me to accompany him and his brother on a rather unusual trip. His brother wanted to consult a chhaya jyotishi in Cambay. The chhaya jyotishi measured your shadow in the noonday sun, consulted his collection of ancient parchments and looked for the one that matched with the measurement of your shadow and predicted the future.

Medora's brother wanted his shadow 'read' because he was keen on getting married and was seeking 'spiritual' advice about whether the young lady he had in mind was the right choice. I found this entire exercise quite ridiculous. I had never had faith nor interest in the 'occult sciences'. I went along with the Medoras because anything was a good change from the monotony of life at Anand.

After Medora's brother got his shadow 'read', they persuaded me to do the same. So as not to appear a spoilsport and also for some fun I stood in the sun while the jyotishi measured my shadow. Shuffling through the bunch of parchment-like leaves, and finding what he was looking for, he read out: 'You have no faith.' I told him he was absolutely right; I was an atheist. Ignoring me, he continued to read out some details about my family and childhood which turned out to be absolutely accurate. He then asked me if he should read me my future. By this time I was rather intrigued so I agreed.

Among the many things the jyotishi told me, a particular detail remained firmly stuck in my mind: 'You are very unhappy in your job right now but within a month you will change it and then you should just sit back and watch,' he read out. 'Your career is set for a phenomenal rise – the kind you can never imagine.' I had smiled sceptically to myself then, but in hindsight what he predicted could not have been truer. Within a month I left the government creamery to join the Kaira Cooperative. The rest, as they say, is history. Till today I have not arrived at any rational explanation for the chhaya jyotishi's prophecies. Certainly it did not turn me into a believer. I continue to have no faith in occult matters and consider this little incident as simply one of life's curious accidents.

I returned to Anand with the Medoras and soon forgot about the prediction. Other, more important developments took precedence. My persistent letters to the Ministry of Agriculture finally paid off and the government decided to accept my resignation. I was delighted. At last, I would escape this godforsaken place. I began to pack my bags joyfully. Just as I was packed and ready to leave, Tribhuvandas arrived at my garage door. 'I hear you are leaving,' he said. 'Have you found another job?' I told him that I had not but I was going to Bombay to look for one.

'In that case, why don't you stay here till you get another job?' he asked. 'You convinced us to order all this expensive equipment which is coming in next week and now you are leaving us in the lurch. None of us here know how to erect all those fancy machines. Why don't you stay here, set up the equipment, get it working, teach our people how to run it and then go?'

Tribhuvandas Patel was a very difficult man to refuse. In the months that I had known him I had come to realise that he was an exceptional person of tremendous integrity and totally committed to the cause of the farmers. I considered it an honour to work with him and agreed to stay on for two months on a salary of Rs 600 a month. Two months, I thought, would be an adequate time to set up the equipment and get the plant working efficiently. Little did I realise that this two-month contract was to be merely the beginning of my lifelong association with Tribhuvandas and the milk cooperatives of Kaira district.

Tribhuvandas gradually involved me deeper and deeper in the workings of the cooperative. He told me that he needed my professional skills to operate and manage the dairy for the farmers and that the farmers would be grateful if I were to stay on at Anand. It was his vision that made him quick to motivate professionals like me to contribute our services to build farmers' cooperatives. It was my fortune, or fate, call it what you may, that Tribhuvandas took a fancy to me and that I recognised an opportunity in what he was asking of me. An opportunity not only to serve myself but also to work for the larger good. This is one of the most important lessons in life that I was to learn. Working with Tribhuvandas and Kaira's dairy farmers, I saw that when you work merely for your own profit, the pleasure is transitory; but if you work for others, there is a deeper sense of fulfilment and if things are handled well, the money, too, is more than adequate.

When I first began working with Tribhuvandas, the farmers and the cooperatives, I had not dreamt in my wildest dreams that this unusual combination of talents and strengths would revolutionise dairying in India. At that critical formative stage of the Kaira Union, Tribhuvandas was the main reason the experiment worked. He was the Chairperson of the cooperative and because he was at the helm, no government or political interference was possible. During those days when the Congress Party ruled without challenge, he was an influential Congress leader in Gujarat. His opinion carried great weight when the party named ministers and members of Parliament from the region. However, so clear and unshakeable were his priorities that if any politician even tried to put as much as a finger on the farmers' milk cooperatives, he would have cut off their entire hand. The rest of us watched and learnt.

In 1950 when I formally joined the KDCMPUL as General Manager, we were a newly-independent nation. Our cooperative was helping to bring economic independence to the dairy farmers of Kaira district. There were so many things to do, so many challenges to overcome, so many opportunities to seize. We felt exhilarated from knowing that what we were doing was important – crucial for our farmers and for the new nation. The greatest satisfaction and joy came from the priceless reward that comes when farmers whose lives depend on your efforts appreciate what is being done for them. For me, helping to build and shape a cooperative owned and commanded by milk producers has always been the greatest reward.

~

I grew with the cooperative and learnt new lessons. One of my first experiences was to learn that within every challenge there is an opportunity.

One of the unique problems with Indian dairying is that buffaloes give double the milk in winter than in summer. In those days, however much milk was produced, we had to send it all to Bombay or it would spoil and be lost. Dara Khurody, the Milk Commissioner of Bombay, was displeased with the fluctuation in supply and decided that we must send the same amount of milk the year round. This is when my legendary run-ins with him began.

I said: 'Mr Khurody, buffaloes give double the milk in winter and I don't know how to plug their udders. I'm afraid you will have to accept all the milk.'

He became extremely angry and retorted: 'But the people of Bombay don't drink one bottle of milk in summer and two bottles in winter. It's your problem, not mine. I cannot take the milk.'

I knew that the Bombay Milk Scheme did not have adequate milk to supply to its consumers and so it imported milk powder from New Zealand and converted it to liquid milk to meet the city's demand. I believed that since there was adequate liquid milk available within the country, the practice was both unnecessary and unfair to our farmers. Never one to shy away from battle, I confronted him: 'Mr Khurody, are you the Milk Commissioner of Bombay or of New Zealand?' I asked. 'Why are you importing milk powder from another country instead of taking milk from our own farmers?'

'How dare you?' he shouted back. 'Who are you to question the government?' And the matter ended there. Khurody refused to take the surplus milk and continued to import milk powder from New Zealand.

It was around this time that I discovered some of the intriguing benefits of 'importing'. For some it meant a trip abroad, for others inflated invoices and other devices about which the less said the better. Suffice it to say that I was unable to stop the import. I complained to the Government of India, too, but all my arguments fell on deaf ears. I began to see then that when the government enters business, the citizens of India get cheated. The greatest repercussion of the government entering into business is that instead of safeguarding people from vested interests, they themselves become the vested interest.

Although we were supplying large quantities of milk to the Bombay Milk Scheme, there was no dearth of obstacles in our way. Once the Milk Scheme officials claimed to have found a fly in the milk we supplied, a huge hue and cry was raised and the media splashed the news. I immediately suspected foul play and suggested to the Milk Commissioner's office that they should do a post-mortem of the fly. I wanted to see if the fly had milk in its lungs – if so, then it had drowned in the milk; but if not, then a dead fly had been put in there to frame our cooperatives. Miraculously, the uproar quietened down and nobody mentioned the hapless fly again.

Another incident that could have created a huge scandal for us was also nipped in the bud. Medora, who worked in Khurody's laboratory in Anand, stayed with Dalaya and me and we were close friends. By this time I had moved from the garage into an old government house where the three of us – bachelors – stayed together. One evening, Medora returned home from work and with an extremely worried look on his face said to me, 'Hey Kurien, we've found formalin in the cooperative milk.' I was shocked. Formalin is a poison and is sometimes used illegally as a preservative since it kills bugs and bacteria. But it is a cumulative poison. We had to investigate and find out its source.

We began the search and finally identified the truck which carried the contaminated milk. We went to each of the six villages from where the truck collected the milk, to find out the source of formalin. We learnt that it came from the milk collected from Pansora village. We went there immediately, called the society's Chairman and searched the collection centre. Sure enough, we found a bottle of formalin. When we enquired about the presence of the bottle at the centre, he explained to us quite gleefully, 'Oh we add this much (indicating a small amount) to every can and then, Sir, there are no rejections!' He was an illiterate farmer and extremely happy that all rejection of their milk had stopped since they started adding formalin. We were horrified and asked him who had advised him to do this. He looked at Medora and said, 'Sir, your chemist came and told me to do this. In fact he even bought the bottle for me.'

Medora was acutely embarrassed. As soon as we returned to Anand he questioned the two chemists who had done this and dismissed them straightaway. He, however, had no authority to dismiss any staff – only the Milk Commissioner could do so and that too after an enquiry and a due official process. Fortunately, Khurody was understanding about it. He reprimanded Medora for acting impulsively and then sent the Deputy Milk Commissioner with a suspension order to be served to the two chemists and to conduct an enquiry. Life at Anand was getting interesting.

One evening Tribhuvandas sent me a message saying that the Deputy Minister from Bombay was arriving by the Gujarat Mail next morning and I should meet him at the Anand railway station. Those days, the job of receiving VIPs had been informally assigned to me, perhaps because I was perceived as a young bachelor who had nothing better to do. When I went to Anand station at five-fifteen the next morning, I learnt that the visitor was Deputy Minister Y.B. Chavan, who later became Chief Minister of Maharashtra, Defence Minister and Deputy Prime Minister of India.

I accompanied the Minister to the dairy to show him the processing of milk. After the tour I took him to my house and as we sat there talking I thought I should tell him about Khurody's repeated attempts to sabotage the progress of Kaira's milk cooperatives. Seeing how upset I became while recounting Khurody's games, Y.B. Chavan took me to the next room – away from the Congress Party workers who had come to meet him – and said, 'You are absolutely right in all that you say. There is nothing I can do right now as I am only a Deputy Minister. But I think it is my duty to tell you that what you are doing is correct and you must always continue to fight for the rights and cause of Kaira's milk producers.'

By 1952, the Kaira District Milk Producers Union Limited was into its fifth year of operation. From a modest 200 litres a day in 1948, milk collection had now reached 20,000 litres per day. The milk collection area was expanding and dispatches of pasteurised milk from Anand to the Bombay Milk Scheme by insulated railway vans were increasing. With the growing demand for milk in Bombay and the potential to procure milk deeper from the hinterland, I realised that very soon we would need a new dairy with enough land to accommodate a rail siding. I knew that as General Manager it was crucial for me to expand my knowledge and hone the skills needed to run a modern dairy. The chance to do so fell into my lap most opportunely.

That same year, two Central Government ministers – Finance Minister C.D. Deshmukh and Agriculture Minister K.M. Munshi – came to Anand to inaugurate the new agriculture institute. Agriculture Secretary Vishnu Sahay accompanied the ministers. Finding appropriate accommodation was still a problem in Anand, I offered to host the Secretary in the spare guest room of my modest bungalow.

Word about the fast growing Kaira Union had obviously reached Delhi, too, and Sahay asked to see how our milk cooperative worked. I took him to some nearby villages and explained to him the Kaira Cooperative's procurement, processing and marketing system. When we returned home, Sahay looked at me and said, 'I am very happy that one of the fellows we sent abroad has come back and done something worthwhile.' He was referring to the Government of India scholarship that had financed my graduate work at Michigan State University.

I deliberated for a moment and then told him, 'You took me from a steel plant and sent me abroad to study dairy engineering when I didn't even know what a cow looked like. Was that any way to select people for specialised overseas training? I learnt nothing of use. I merely had a good time. So don't take credit for what I have done, or for the fact that you sent me abroad.'

'That's very funny,' smiled Sahay.

'Not at all,' I replied. 'It's very tragic.'

'What should we have done then?' asked Sahay.

'You should have taken a fellow like me now, who's worked for three or four years in a dairy and has picked up some knowledge and knows what to look for,' I said. 'If you can send me out now I can learn something really worthwhile.'

'Well, if that's the right thing for you to do, why don't you go somewhere?' Sahay asked.

'Go how? And where is the money?' I said.

And our conversation ended there. It was probably then that Vishnu Sahay got the idea of sending me to New Zealand on a senior fellowship under the Colombo Plan. My fellowship lasted from October 1952 to April 1953. I spent the first five months in New Zealand, the haven of dairying, and another two months in Australia. In New Zealand, I was attached to Prof. Kelvin Scott, head of the dairy engineering division of the New Zealand Dairy Research Institute at Massey Agricultural College. When Prof. Scott heard that I had earlier studied under Prof. A.W. Farrall at Michigan State University, he insisted that I share his office and desk, treating me as an equal rather than his student.

New Zealand, I discovered, was a charming country full of very fine people. Hardly surprising, since most people in New Zealand were farmers and farmers are good-natured people. I had been provided a beautiful Morris Minor. On one occasion a policeman stopped me and asked to see my driving licence. When I told him that my driving licence was back in India he simply asked me to demonstrate how to reverse the car and allowed me to go. Prof. Scott and his wife were wonderful to me. Their house was always open to me and I can never forget the many occasions when I enjoyed their warm hospitality. Their two-year-old daughter was a delight and always the highlight of my visit to their home.

New Zealand, I soon learnt, had no private dairies. The New Zealand Cooperative Dairy Company managed all dairy plants. Each unit manufactured a specific product. Part of my training obliged me to visit each plant in turn and study its production process. I must have visited a hundred dairy plants. I made it a point to speak to every class of worker, even the truck drivers who transported milk and its byproducts between dairy plants. I learnt about heat balance sheets, heat transfers, waste management and everything that goes into the successful making of milk powder. The most valuable part was that I had to learn about manufacture of milk powder from cow's milk even as I would have to come back to India and organise efforts to make milk powder from buffalo's milk, since most of our milk came from buffaloes, not cows.

In the early fifties, manufacture of milk powder from buffalo milk was considered impossible, although we had actually done it on a small scale in our research creamery. To add to our concern in Kaira, reputed dairy technologists such as Prof. William Ridett of the New Zealand Dairy Research Institute, New Zealand, and Prof. H. D. Kay of the National Institute for Research and Dairying, Reading, UK, upheld this view. However, my training at New Zealand made me confident that it was, indeed, possible to produce milk powder from buffalo milk even on a commercial basis. On my return I knew I would have to persuade my friend and colleague Dalaya to see how this could be done. This was a breakthrough that was unimaginable at that time.

My five fascinating months in New Zealand and Australia ended with an offer from the Government of New Zealand for aid to develop dairying in India, especially in Kaira district. New Zealand gave 50 per cent of the promised assistance as a gift to the Kaira Union and the other 50 per cent as a loan to the Government of India. Needless to say that this gesture from the New Zealand government formed part of a dream to which the interacting faith of a great many people finally gave substance. The offer of aid and assistance from New Zealand contributed significantly to the struggle of the farmers of Kaira district. Of course, by providing me training and orientation in their country, New Zealand also helped rural producers of Kaira district to strengthen their partnership with professionals – to ensure that modern technology can also benefit our villages.

The Kaira Cooperative grew from strength to strength. We worked incessantly and managed to turn it around until it started making profit. At this point I voiced an idea which had been germinating in my mind for a while. I told Tribhuvandas that we would need another modern dairy and we should try and get some land for it. I calculated that this new plant would cost some Rs 40 lakh. He was stunned and said it would be very difficult to raise such a huge amount. Once more I played my cards right. To emphasise how important I believed the new dairy was, I told Tribhuvandas that if the cooperative did not agree, I would have to leave and look for a new job. As he had the first time I had thought of leaving, Tribhuvandas persuaded me to stay and somehow managed to raise the money for the new dairy. He had the foresight to realise that the dairy could not be run without professionals.

Tribhuvandas Patel was a man of extraordinary ability, leadership and integrity. Yet he recognised that his strengths needed support and had to be complemented by the skills of professionals. He heeded and respected my advice as a professional – just as he expected me to heed and respect his views when it came to the cooperative side of business. He knew Kaira's farmers and how they thought. What is important is that although he was one of Gujarat's most important political leaders, he never interfered with the operations of the cooperative. He supported me and my team, helping to ensure that we made our best contribution to the cooperative's success.

I cannot emphasise enough that our 'miracle' at Anand would not have been possible had Harichand Dalaya not been on our team. I knew that the Kaira Union had tremendous potential but I was also aware that there was a lot I did not know about dairying. Dalaya had studied with me in the US and during one of my visits to Bombay, over a long lunch, I persuaded him to join me at Anand. He was one of the greatest dairymen I have met. A Yadav from Uttar Pradesh, Dalaya took great pride in tracing his lineage back to the gopis of Lord Krishna! He was born and brought up among the milk producers; his family once owned a large dairy in Karachi. Unfortunately, Dalaya was forced to leave for India when the Partition took place. The dairy, with three hundred magnificent red Sindhi cows, had to be closed down. Dalaya was not just a dairyman. He was a brilliant dairy technologist as well. When he joined us, I told him that our division of labour would be very simple: I had the gift of the gab, he had the expert knowledge of dairying; I would do the talking, he would do the work! He must have approved of this because he stayed with us in Anand for thirty-five years.

Tribhuvandas, Dalaya and I soon came to be known as the Kaira Cooperative's 'triumvirate'. Each one of us was distinctly different in manners and skills. Perhaps an organisation needs a mix at its helm. Together we were able to propel the cooperative forward rapidly. It is not that we never disagreed. We did, but each had such a tremendous respect for the integrity and the strength of the other two that all disagreements were resolved privately and maturely without allowing our shared vision to ever swerve from the larger cause.

Another person who came into my life in Anand, and touched me and my family, was Maniben Patel, Sardar Vallabhbhai Patel's daughter. Maniben visited Anand and the cooperative regularly. She was never formally part of the Kaira Cooperative; she had never held an office. She kept in touch only because she was Sardar Patel's daughter and she knew how concerned her father had been about the welfare and progress of the dairy farmers of Kaira. She was always welcome at Anand, not just as Sardar Patel's daughter but because of her own qualities as a human being. Although she was a rather formidable woman and had a crusty exterior I discovered that she had a very soft heart and we became extremely good friends.

Maniben was a woman of tremendous honesty and loyalty. She had dedicated her entire life to her father. She told me that when Sardar Patel passed away, she picked up a book and a bag that belonged to him and went to meet Jawaharlal Nehru in Delhi. She handed them to Nehru, telling him that her father had instructed her that when he died she should give these items to Nehru and no one else. The bag contained Rs 35 lakh that belonged to the Congress Party and the book was the party's book of accounts. Nehru took them and thanked her. Maniben waited expectantly, hoping he would say something more, but he did not, so she got up and left.

I asked her what she had expected Nehru to say to her. 'I thought he might ask me how I would manage now, or at least ask if there was anything he could do to help me. But he never asked,' she explained. She was extremely disheartened and in a way the incident revealed the extent of strain in the Nehru-Sardar Patel relationship. It was quite distressing to see that neither Nehru nor any of the other national leaders of the Congress Party ever bothered to find out what happened to Maniben after her father died.

She did not have any money of her own. After Sardar Patel died the Birlas asked her to stay at the Birla House for a while, but the arrangement did not suit her so she left to stay in her cousin's house in Ahmedabad. She had no car, so she travelled in buses or by third class in trains. Later, Tribhuvandas helped her to get elected as a member of Parliament and so she got a first-class pass but, like a true Gandhian, she continued to travel only third class. She wore only khadi saris made out of thread she had spun herself and wherever she went she carried her spinning wheel.

Rather predictably, after Sardar Patel died Maniben transferred her father's fixation to Morarji Desai who, unfortunately, did not have any time for her. She would accompany me whenever I went to meet him and he would make her wait in another room while he called me into his bedroom. I used to find his behaviour quite upsetting and once I raised it with him. I told him that his indifference to Maniben bothered me. I explained to him that she had come all the way to see him and instead of behaving so curtly, a smile and a kind word from him would make her day. He never meant to be rude but somehow he treated her with indifference.

After all the sacrifices that Sardar Patel made for the nation, it was very sad that the nation did nothing for his daughter. In her later years, when her eyesight weakened, she would walk unaided down the streets of Ahmedabad, often stumble and fall until some passerby helped her up. When she was dying, the Chief Minister of Gujarat, Chimanbhai Patel, came to her bedside with a photographer. He stood behind her bed and instructed him to take a picture. The photograph was published in all the newspapers the next day. With a little effort they could so easily have made her last years comfortable.

~

When I returned from New Zealand in April 1953, my mind was bursting with plans for our new dairy and to begin the production of milk powder. But fate had planned a happy detour along that path. For a while now, my family – especially my mother – had been unhappy about my bachelor status and had been suggesting that I get married and 'settle down'. I was extremely wary of the practice of 'arranged' marriages and had managed to resist all attempts of my well-meaning family to trap me into matrimony. However, this time my eldest brother called me and told me very firmly that I must go to my mother's home in Trichur to meet the young lady whom he had already met on my behalf. If I did not hurry, he hinted, the lady would not wait for me. This 'young lady' was Susan Molly Peter, whose parents were extremely close friends of my family.

Our families went back a long way. My father and Molly's were both doctors and had studied together. The families met often and knew each other very well. It was quite amazing that despite our families being so close, Molly and I had never set eyes on each other. It was with a great deal of trepidation, therefore, that I headed for Trichur towards the end of May 1953, almost certain that I would return to Anand still a bachelor. I must have been uncharacteristically nervous when I met Molly for the first time because she later told me that I spent the entire time with her talking about Maniben! Suffice it to say that after my first meeting with Susan Molly Peter I informed both the families that I was not going back without marrying her. And what is more, the wedding had to take place within fifteen days since both Dalaya and Tribhuvandas were scheduled to go abroad and my quick return to Anand was unavoidable. It must have created an organisational nightmare for both the families, but somehow everything went off without a hitch.

The only practical detail I recall about the wedding was when my aunt, Mrs John Matthai who had come for the event, asked me what I planned to wear for the ceremony. She was horrified when I showed her my old suit, which I had purchased when I went to the US. She immediately called for her tailor to take my measurements and had a silk kurta and dhoti made for me. Molly and I were married on 15 June in the All Saint's C.S.I Church at Trichur's Mission Colony.

The ceremony took place at ten a.m., and at four p.m. the same day we were on the train for Bombay where we had to meet Tribhuvandas and Dalaya, who were on their way to catch their flight out of the country. After that we boarded the train from Bombay, which reached Anand at four-thirty in the morning. The wedding had been such a whirlwind affair that I had no time to inform my colleagues at Anand. As a result there was nobody at the station to receive my bride and me. A quiet homecoming. My house was literally across the road from the station and so Molly and I picked up our bags and walked home. 'Home' was a modest house with a bedroom, a drawing room, a kitchen and a guest room, which was then occupied by an official visitor from Denmark.

In hindsight I have often wondered how Molly managed so gracefully during those early years of our marriage. With both Tribhuvandas and Dalaya away, I dived headlong into work and was almost completely immersed in the problems facing the fast-growing dairy. There was very little time for the luxury of personal adjustments. I worked late and travelled a great deal and Molly had to fend for herself. Moreover, until 1958 the dairy did not have a guest house. Since my house was seen as the unofficial guest house for Kaira Cooperative, there was an endless stream of visitors and house guests who had to be looked after and provided meals.

I could not have been an easy person to live with during that period and I still recall how on occasions, during our little disagreements, I would tell Molly that I should never have married because at least then she would not be occupying space in my house which was so desperately needed to host more official visitors to the cooperative! But Molly, with her poised and serene manner and an enviable inner strength, put up with my impatient and idiosyncratic ways.

She did have support from some of our extraordinary friends in Anand. To begin with, there was the inimitable Maniben. Once Maniben realised that Molly did not conform to her preconceived, stereotypical notion of a Christian woman – westernised, frock-clad and forward – she was thrilled. She took Molly under her wing. She had always been fond of me but after I got married, Maniben became almost part of our family, a surrogate mother to us. One of the first bits of advice she gave Molly was: 'You must always remember that you are Kurien's second wife. His first wife is the dairy. Don't ever forget that and don't make yourself miserable by being jealous. And never, never try to snatch your husband away from his first wife.'

Besides Maniben, our friends R.H Variava and his wife Zarine were exceptionally kind and caring towards us. Variava, who had worked with Polson dairy since 1939 as Deputy Manager, was now the Manager there. Despite the fact that he was an officer with our rival dairy, Variava had great regard for me and for the work that Kaira's Cooperatives were doing. The Variavas were quintessentially Parsi – warm, large-hearted, selfless and giving. It was because of people like them that Molly did not feel like an outsider, cut off from her own people, in this little town. Like Maniben, they too loved her like a daughter and often, when I left Anand on my travels, they would whisk her off to stay with them so that she did not feel lonely. Zarine Variava was a good-natured and friendly person who frequently and sportingly bore the brunt of my practical jokes. The Variavas kept poultry. This provided an incentive for me to bet with Zarine on something or the other, provided of course that I was certain to win. The prize was always a chicken and as a consequence, our dining table frequently boasted of chicken curry which the Variavas happily shared with us.

The Kaira Cooperative continued to expand. In order to tackle the problem of surplus milk in winter we concretised our decision to manufacture our own milk powder. Dara Khurody may have spurned our surplus milk, but to give him credit, he did bring to Anand representatives of the United Nations Children's Fund (UNICEF). When they saw the predicament we were in, they promised to donate to the dairy equipment for making milk powder. Hearing about this magnanimous UNICEF offer, a supremely confident Khurody declared: 'Powder cannot be made from buffalo milk. I have a letter from the great dairy expert, Prof. William Ridett of New Zealand, who says it is technically impossible.'

I was to learn yet another valuable but sad lesson: that the technical advice of 'experts' is all too often dictated by the economic interests of the advanced countries and not by the needs or ground realities in developing countries. Without exception, technical experts from England and New Zealand told us that buffalo milk could not be converted to milk powder.

We showed them how it could be done.

~

In 1956 the UNICEF representative, T. Glen Davies, came to India. Donald Sabin of UNICEF's Food Conservation Division in New York was his senior. As they had indicated to us, UNICEF was keen to provide assistance to help us set up a milk powder plant in India. The terms of assistance required free distribution of milk to children and expectant mothers valued at one-and-a-half times the cost of the plant, to be repaid over a period of five years.

Jivaraj Mehta, the then Chief Minister of Bombay state (which then consisted of Maharashtra and Gujarat), called me for a meeting as the General Manager of KDCMPUL, and told me that I should accept this plant for the Kaira Cooperative Union. I was a little apprehensive since it would mean paying the additional 50 per cent of the value of the equipment donated. I suggested that perhaps the Government of Bombay might want to help in repaying the amount. But Jivaraj Mehta explained to me about the moratorium on repayment for five years and I agreed. Tribhuvandas during this time was on a trip abroad. Aware of this, the Chief Minister said, 'Kurien, you are only a Manager. Your Chairman is out of the country. You should call a board meeting and ask them to approve this. In fact, better still, call a general body meeting of KDCMPUL and ask them for approval.' He also promised to ask the Deputy Minister from Kaira, Babubhai Jessubhai Patel, to speak to the district leaders to facilitate the entire process. The Kaira Union's general body approved and accepted UNICEF's proposal.

However, one major hurdle remained. The final approval for the assistance had to be obtained from the Milk Commissioner of Bombay state. We were aware of Khurody's thoughts on the matter. He was convinced that the Kaira Cooperative could never make milk powder from buffalo milk and, in fact, had produced letters from experts to support his claim. In the face of Khurody's opposition, I feared we would never get the UNICEF assistance. It was time, I realised, to take matters into our own hands.

Dalaya and I decided to go to Bombay to meet Dinkarrao Desai, Minister in charge of dairying. We had one meeting in the morning with the Minister during which Khurody persistently stonewalled the proposal. We were unable to arrive at any agreement so we decided to meet again after lunch to try and thrash out matters. Dalaya and I needed an urgent break from the tense atmosphere in the Minister's room so we decided to go out for lunch. As we were driving to a restaurant Dalaya suddenly exclaimed, 'Hold on. There is a Larsen & Toubro powder plant somewhere here in Bombay where we can demonstrate that milk powder can be made from buffalo milk. This can prove to Khurody and Desai that it is possible.'

In our excitement we forgot the lunch and headed instead to the L&T office. Here we were told that the machine we were looking for had already been sold to Teddington Chemicals, a British laboratory company in Andheri in Bombay. I asked the L&T manager, Axil Petersen, 'Would you be interested in an order from us for setting up a large milk powder plant?' Petersen replied that they certainly would be. 'Then please speak to the Andheri laboratory and ask them to lend us their powder plant for a demonstration tomorrow.' Having made what Petersen must have thought a very strange request, we returned to our meeting with the Minister.

Petersen made the necessary arrangements with Teddington Chemicals and telephoned Dalaya even as we were in the middle of our meeting with Desai, Khurody and UNICEF representatives Sabin and Davies that afternoon. When Dalaya told me that everything had been organised, I announced at the meeting that at nine a.m. the next day, at Teddington Chemicals' laboratory in Andheri, we would demonstrate how to make milk powder from buffalo milk. They were astounded. I then asked Khurody to organise for some skimmed buffalo milk to be brought to the laboratory the next morning. He agreed to do so reluctantly.

The next day we met at the Andheri laboratory where Dalaya and I proceeded to convert the buffalo milk into milk powder. Khurody was still sceptical. 'But what about its solubility?' he demanded. I was prepared for this question. I simply added the milk powder to a beaker of distilled water and it dissolved completely. 'But what about its taste?' Khurody persisted. I promptly offered the reconstituted milk to Dinkarrao Desai who sipped it and said it tasted absolutely fine.

'It has now been proved that milk powder can be made from buffalo milk. UNICEF will assist Kaira Union to set up the powder plant,' an excited Sabin announced.

'But ... ,' began Khurody, trying to introduce yet another doubt.

'No more "buts" please,' Davies said, cutting him short. Then pointing to Dalaya and me he remarked, 'I like the cut of the faces of these two young lads and we are, indeed, going ahead with the proposal.'

The Bombay state government finally approved the project. Dalaya and I returned to Anand victorious.

It was only much later that Dalaya confided to me that it was a blessing that Khurody did not seem to know too much about the technical aspect of dairying. He explained that making milk powder from skimmed buffalo milk is not very different from making powder out of skimmed cow milk. The problem can occur in the condensing unit, where unskimmed buffalo milk may curdle. But since the condensing unit operation was not part of our demonstration in Bombay, Khurody was none the wiser.

'And now how will you solve that problem?' I asked Dalaya, a trifle apprehensive.

'Don't worry. Even that can be resolved by adjusting the balance of the salts in the milk that is fed into the condensing unit,' he explained.

The project got underway. UNICEF sent us a cable stating that following their tender procedure they had decided to donate a Dutch Volma milk powder plant to the Kaira Union. I was a little worried about this. I called Dalaya and others for a discussion. We realised that it was imperative that the powder plant supplier should have an Indian office to provide ongoing support and service. I sent a cable to UNICEF saying: 'Kaira Union does not want a Volma powder plant but a Larsen & Toubro Niro powder plant.'

Evidently, UNICEF was unhappy about this and retorted with a curt cable: 'UNICEF is not accustomed to being told which powder plant to supply. The Kaira Union risks jeopardising UNICEF assistance.'

I could not back out at this stage so I called up H.M. Patel, who besides being a friend and supporter of the Kaira Cooperative Union, was also at that time the Principal Financial Secretary to the Government of India. As always, he heard me patiently. I explained the situation to him and sought the Government of India's backing to take on UNICEF. H.M. Patel assured me of the government's support and also backed me personally. With that support, I sent another cable to UNICEF saying: 'Kaira Union is not accustomed to being told what it should have. Willing to jeopardise project if UNICEF does not give Larsen & Toubro Niro plant.'

We learnt another useful lesson: with adequate support, confrontation at the right time pays off. UNICEF agreed and the Kaira Cooperative Union got its L&T powder plant.

The Kaira Cooperative's old dairy was right in front of the

railway station in an extremely congested area. As we searched for a suitable site for the new dairy, our omniscient Barot Kaka informed us that Khurody had already got land for a government dairy in Anand. After some long and occasionally heated discussions with Dinkarrao Desai we managed to get this land for the Kaira Cooperative. The new dairy was built there.

We decided to invite the President of India, Rajendra Prasad, to lay the foundation stone for the Kaira Cooperative Union's new dairy. It was an important occasion. For the first time in the history of the nation, farmers would actually own the country's most technologically advanced dairy. The realisation of farmers' dreams rested on the plant's capacity to manufacture powder from surplus milk.

On 15 November 1954 the train carrying the President came into the railway siding at the dairy. He was taken straight to the site and the customary ceremonies began. First the bhoomi pooja was performed. The highlight of the event was when the President was about to place the first foundation stone in the earth, suddenly, to everybody's amazement a mouse came scampering by and jumped over the stone. The entire gathering was overcome with joy because the mouse is seen as Lord Ganesha's vehicle and, therefore, considered auspicious.

~

Once the dairy's foundation stone was laid by the President of India, Maniben, who was a Congress Member of Parliament at that time, asked me, 'When will your dairy be ready?' I told her that it would be ready in about a year's time. She asked me who I had in mind for its inauguration and I replied, 'Jawaharlal Nehru, of course.' She said: 'Well if you want him, I better talk to him soon and book him for that day.' She spoke to the Prime Minister and informed me that he had agreed to inaugurate the dairy on 31 October – Sardar Vallabhbhai Patel's birthday.

A bit nervous, I called Dalaya and told him that the date for the formal inauguration had been fixed. 'We are building India's first milk powder plant. We've been told that powder cannot be made from buffalo milk. And here we are fixing the date for the dairy's opening. How are we to build this dairy in eleven months?'

But Dalaya was confident and told me that we should not back out at this moment and that the dairy would, indeed, be ready in time. So I spoke to Maniben, confirming the date. It was decided that the Prime Minister would arrive at Anand on the appointed date, inaugurate the plant and dedicate it to the memory of Sardar Patel. We had the President to lay the foundation stone and the Prime Minister to inaugurate the dairy. A winning combination.

Things had to move fast. With the inauguration date already fixed, we were committed to setting up the new plant in record time and we had our share of problems.

As the construction work began we started getting some thirty to forty wagonloads of material each day, which had to be kept at the siding. I was informed that the Station Master was demanding Rs 5 per wagon for doing this. I was livid and immediately telephoned F. J. Heredia, who was then Kaira's Collector and District Magistrate, demanding some action. Heredia asked me to calm down. He came over himself and took me with him to meet the Station Master. Heredia said to him, 'I am the Collector and I have come here to thank you.' I looked at him in amazement but Heredia continued, 'I have come here to thank you for all that you are doing to help build India's first milk powder plant. You must be very proud that you are able to play a role. So I have come on behalf of the Government of Bombay state to thank you. And I would also like to tell you that in my report, I am going to mention all the things you are doing to help the work of the dairy.'

I couldn't believe my ears. 'What are you doing?' I asked him. 'I asked you to take action against him.'

'Yes, yes, I understand,' explained Heredia. 'But the most I can do is to transfer him. Then another station master will come. He will also be like him. There will be no difference. So what have we achieved? Now we have, in good sense, sounded him a warning. He will behave properly.' And indeed, the technique worked and the Station Master became a helpful friend.

Then, just when things appeared to be moving smoothly, Molly and I were sitting down to lunch one day when a messenger came running and informed us that there had been an accident in the boiler house. I rushed to the dairy. Somehow, Khurody too happened to be there that day and he arrived at the scene of the disaster as well. The roof of the boiler room, which was being concreted, had collapsed. It was a horrifying sight. We had no idea how many workers were buried under that huge pile of concrete.

Khurody, who was watching all this from a distance, remarked caustically, 'Naturally, when you people build pillars like matchsticks, what do you expect?' I ignored him and told the mechanic to bring a bulldozer quickly and pull out the beam. The mechanic began dithering, saying that it was dangerous and if they did that the wall would fall over them. So I told him to move out of the bulldozer and I operated the bulldozer and pulled the beam out myself.

Once that was done, with the help of some workers, we removed the wet concrete before it solidified. Fortunately, not a single person was seriously injured. The only thing that happened was that three stitches, which I had in my bottom after a minor surgery, broke, and to everybody's horror, I had a patch of blood on my pants!

Another hurdle cropped up when Larsen & Toubro sent a team of Danish engineers to help. Trouble further increased when it was brought to my notice that these engineers were not paying heed to Dalaya's suggestions. I called them and gave them a warning, 'Please remember that when Dalaya speaks, he is speaking for me, he is speaking for the owners who are paying your salary and you must carry out his orders. Who are you to question Dalaya? His judgement is final and you will do it his way.'

I also telephoned Toubro and called him to Anand. He came the very next morning and promised to talk to his engineers and get back to me.

When he finally came to meet me, Dalaya was with me. Toubro said, 'From this moment onwards, I hold myself responsible to you and to Dalaya. Every week I will give you a report of our progress.' After that we had no cause to complain. Work proceeded steadily.

Finally, I set off to Bombay to order the two boilers for the dairy from a firm called J.N. Marshall. I learnt that the ship carrying the boilers would only reach Bombay in September, a month before the inauguration. I knew the process of berthing and offloading could take time so I approached Morarji Desai, who was then the Chief Minister of Bombay state, requesting him to ensure that the ship carrying the boilers was given priority for berthing. Morarjibhai promised to do whatever he could and he called the Chairman of the Bombay Port Trust. The Chairman, however, refused, saying that only the board of the Bombay Port Trust could take such a decision. Morarjibhai asked him to convene a board meeting the very next day.

The Bombay Port Trust Board cast the vote in our favour and the ship was given priority. But the race was still not over. I spoke to a friend in the Railway Board and asked him to instruct the driver of the goods train carrying the boilers to travel non-stop from Bombay Port to Anand. He was shocked at my request and asked, 'But what about the other goods being carried to intermediate destinations like Surat and Baroda?' I convinced him that those goods could be offloaded on the train's return journey. After many hours of tense negotiations, the boilers finally arrived at Anand station, barely fifteen days before the plant was to be inaugurated.

Meanwhile Marshall's son-in-law, Darius Forbes, and four

other engineers were already in Anand to set up the boilers. These young engineers had hoped that the experience at Anand would actually be a practical lesson for them and they would learn how to erect boilers. They were shocked to learn that they were expected to set up and commission the boilers within the fortnight. They made a panic call to Marshall in Bombay, who quickly summoned the best erection and commissioning engineers from his Principals in Scotland. Finally, difficult as it is to believe, the boilers were fired with barely a day to go for the inauguration.

The boilers began generating steam and the first batch of milk powder rolled out in a trial run of the new plant, past midnight. So thrilled were we with the success that all of us present there indulged in a spontaneous milk-powder fight! Just four hours later, Prime Minister Jawaharlal Nehru was scheduled to inaugurate the plant.

Our friend and well-wisher, Chief Minister Morarji Desai, had already arrived in Anand four days before the inauguration. Clearly, he had been 'briefed' by Khurody in Bombay, for he took me aside and said anxiously, 'Kurien, I am told the plant will not be ready in time. How could you do this? I have invited the Prime Minister and now what will I tell him? Khurody says nobody has built and commissioned a powder plant in one year.'

I tried to put his mind at rest and assured him that the plant would, indeed, be ready for inauguration.

'And just in case Khurody is right? What will you do then?' countered a worried Morarjibhai.

'Sir, finally, what is to be shown to the PM is powder rolling out of the hopper,' I said. 'If need be, some bags of milk powder will be kept on the next floor and fed into the hopper from an inlet upstairs, even as the PM is being shown around.'

Morarjibhai quietly walked away. His silence was a consent to this little trick we had up our sleeve, should it prove necessary.

But on the appointed day – 31 October 1955 – we had no need to cheat.

Jawaharlal Nehru arrived at Anand to tremendous fanfare. His daughter, Indira Gandhi, accompanied him. We first took them to my house. We had already received detailed instructions about his breakfast, and how he liked his coffee and milk piping hot. However, we were faced with the slight problem of getting the right trademark rose for his buttonhole. It had to be got in advance but had to be kept at the right temperature because it must look absolutely fresh. It had to be the right shade of red, the right size and in just the right degree of bloom. Molly had to experiment with many a rose. She finally figured out that we would have to store the flower in the fridge for a certain time and then keep it at room temperature for a certain time before offering it to the Prime Minister. We did all this and had the rose all ready for him. To our surprise, when Jawaharlal Nehru came out of the bedroom, he already had a rose in his buttonhole, picked from the flower vase in his bedroom. Then he saw us with the rose on a platter. He immediately removed the one from his buttonhole and put on the one Molly offered him.

Jawaharlal Nehru and Indira Gandhi were taken on a guided tour of the complex. Our machines seamlessly rolled out milk powder made from buffalo milk. Of course, there were some hitches, but nothing that we could not handle. For instance, Darius Forbes and his group of engineers were eager to be introduced to the Indian Prime Minister. When I took Nehru to the boiler house they were so excited that instead of returning to the boiler house to stoke the boiler and monitor the steam flow, they began following the VIP entourage around. By the time we entered the dairy, the overheated milk was frothing in the vessels. Fortunately the Prime Minister did not know that it was not supposed to froth and the engineers returned to the boiler room in time to prevent further mishaps.

Khurody proved to be another nuisance. Every time I began explaining to Nehru about the dairy, he would butt in. He continued to interfere until Morarji Desai finally noticed this and pulled him back by his shirt.

Unfortunately, while the formal introductions were being made to the Prime Minister, due to an oversight, Toubro was left out. Therefore after the function I invited him to my house where Jawaharlal Nehru and Indira Gandhi were our guests for the day. At home he remarked cheerfully to Molly, 'Mrs Kurien, Cleopatra used to bathe in ass's milk. You can now bathe in buffalo milk!'

I went to him, taking Toubro with me and said: 'Sir, I have kept to the last the man to whom I owe the most. This is Mr Toubro.' The Prime Minister and Toubro had a long chat. Toubro had brought his own photographer along – an extremely well-trained fellow. Many memorable photographs were taken while Toubro chatted with the Prime Minister.

It all ended well. Certainly, Tribhuvandas, Dalaya and I had taken some calculated risks and they had paid off. Free India's first Prime Minister inaugurated the nation's first modern dairy products' plant and gave an encouraging speech praising the Kaira Cooperative for its initiative and courage. As the Prime Minister was leaving the premises, Morarjibhai said to him, 'Mr Kurien has not just built and commissioned this dairy in record time, but this is the first milk powder plant in the world that makes milk powder from buffalo milk.'

It was one of the moments that I have treasured all my life. Jawaharlal Nehru turned to me, embraced me and said, 'Kurien, I'm so glad that our country has people like you – people who will go ahead and achieve even that which seems unachievable.'

We can never forget, however, that the Kaira District Cooperative Milk Producers Union could not have built this new dairy without generous help from others. UNICEF gifted us dairy machinery worth Rs 800,000 and in return for this we had to distribute Rs 12 lakh worth of milk or milk powder free to children and expectant mothers. The Government of New Zealand donated equipment worth Rs 300,000 under the Colombo Plan as well as the services of one of their best engineers to help install the equipment. And of course, the Government of Bombay gave a loan of Rs 1,000,000. The rest of the money (the dairy ultimately cost Rs 48 lakh, exceeding by Rs 8 lakh the budgeted sum) came from the Kaira Cooperative's own funds. It was the largest dairy in all of Asia. Sardar Patel's dream of the farmers owning their dairy had finally been realised.

Those days I was in the habit of taking a walk around the dairy just to ensure that all was moving well and efficiently. One day while I was on one of my unannounced rounds in the cold store, I spotted an old employee, with a big moustache and a beard, who had opened the lid of one of the milk cans and was sucking the cream. Suddenly, he looked up and saw me. We stared at each other for a moment. There was cream dripping from his mouth, onto his chin and he faltered, 'No, no Saheb, I am not drinking, I am not drinking.' I just turned around and walked away. But the very next day, I told the Manager that every worker had to be given half a litre of milk. These men were handling vast quantities of milk all day long and they were hungry. It was not fair that they did not have a share of the milk.

~

It was during this period that my wife's brother-in-law, K.M.Philip, who ran his own company, Philips Tea and Coffee, urged me to start thinking seriously about the finer points of marketing Kaira Cooperative's products. Over many an involved discussion at his house in South Bombay, I was initiated into the minutiae of running a business – details such as branding, distribution and the need to retain an advertising agency. I returned to Anand and spoke to my colleagues about all this and, to begin with, we tried to find an appropriate brand name.

Several heads were put together and during an intense

brainstorming session, a chemist in our laboratory suggested, 'Why not "Amul"?' It seemed to be just what we were looking for in terms of portraying the image and the ideals behind our cooperative venture. The word came from the Sanskrit word 'amulya' which means 'priceless' and denoted and symbolised the pride of swadeshi production. It was also short, catchy and it could, rather effectively, be used as an acronym for Anand Milk Union Limited – certainly easier on the tongue than Kaira District Cooperative Milk Producers' Union Limited. It met with complete approval and 'Amul' came to stay. In 1957 Kaira Cooperative registered the brand 'Amul' – a word that would soon become a household name.

By this time I had realised that if we wanted to hold our own in the market, the professional services of an advertising agency were vital. K.M. Philip also very astutely urged us to leave the business of advertising to the professionals. He suggested a few names. Among the names he gave us were J. Walter Thompson, Grant Advertising and Press Syndicate. As we wanted an Indian company we finally settled for Press Syndicate.

Not unexpectedly, I encountered some resistance from Tribhuvandas. He did not quite understand advertising. Unused to the marketing world, he was reluctant to commit money for an ad campaign for Amul butter. I wanted Rs 2 lakh to start with and he was quite shocked. 'Rs 2 lakh? For advertising?' he asked, disbelievingly. But I insisted and, therefore, he relented.

It was with butter that the Kaira Union started its career in the world of brands and Jit Kantawala who looked after the Amul butter account at Press Syndicate did an exemplary job. He managed to give Amul just the image we had in mind – that of a precious, 'priceless' product that the consumer could trust completely.

I was getting involved deeper and deeper into the working of the cooperative. In the course of reading and studying what cooperatives were and what they stood for I became convinced that the cooperative structure is tailor-made for the dairy industry. It was little wonder then that in the US, the capital of capitalism, 85 per cent of the dairy industry was cooperative. In New Zealand, Denmark and Holland, 100 per cent and in erstwhile West Germany 95 per cent of the dairy industry was cooperative. So with Amul we had proved the point that it was applicable in India too.

The more I worked with Kaira's farmers and the more I saw the hard life they and their families led, the more committed I became to their cause. As the Chief Executive of their cooperative, my main goal became to ensure the best deal for the farmers, within my capacity, without exploiting the consumer. The best way to do this was, of course, to give the consumer products of extremely high quality and that is what we at Amul worked hard to do.

Very soon I was convinced that one of our key areas of concentration would have to be marketing of these products. One of the earliest lessons I had learnt was that Amul existed because, barely a few hundred kilometres away, Bombay existed. There could be production here only because a market was there. There could have been no production of anything unless it was marketed at a price advantageous to those who produced it, which provided them with an incentive to produce more and more. Indeed, there would have been no Anand if there were no Bombay. For this reason alone I would pay my tribute to the Bombay Milk Scheme.

From the very beginning I was convinced that a cooperative, too, must be a business enterprise and it has to run as a business enterprise. If a cooperative forgets this, it will fail; it will collapse. I ensured from the start, therefore, that Amul always operated as a business enterprise – but at all times keeping in mind that the business was to maximise the price paid for the milk, not in order to maximise the dividend, as is the case in the private sector. We did this by manufacturing value-added products which allowed us to give farmers a higher milk price every year.

From liquid milk we went into the production of milk powder and very soon into butter. Until then, the manufacture of butter had been almost totally monopolised by Polson. Very soon, with modern technology at our disposal, we broke into the butter market and soon Amul butter became a roaring success. Sadly for Polson, however, Amul's triumph in butter presaged their doom.

One day after Kaira Cooperative's milk powder plant was up and running full speed ahead, we were informed that the Chief Minister of Bombay state wanted to visit the dairy and he would arrive at Anand early next morning by Gujarat Mail. The Chief Minister was none other than Y.B. Chavan, who had earlier visited Anand as a Deputy Minister. Once he had seen the plant, he asked me if we could go to some quiet place to talk. Leaving the rest of his entourage behind, I took him to my residence.

'I remember, Kurien, when I visited Anand as a Deputy Minister, you had confided to me that you were having a lot of problems with the government,' he said. 'At that time I was unable to help you. I am the Chief Minister now. So tell me all your problems and I will solve them immediately.'

I was very touched that he had remembered but I answered him in all honesty: 'I have no problems now, Sir.'

Chavan looked rather amazed and wondered what had happened to all my troubles with Khurody. But Khurody was far smarter than I had anticipated. Only two days before the Chief Minister was to visit Anand, he called me from Bombay and said that our battles had gone on for long and it was time to bury the hatchet. 'Kurien, let us be friends,' he said to me. Not one to let such an opportunity pass by, I asked him to make a commitment that he would stop messing around with the Kaira Cooperative and instead, would help us to secure the growth and future of the cooperative. He agreed and from that time onwards we got onto a totally different footing.

With the passage of time, I realised that one could not blame Khurody and his colleagues for their attitude. In India, all along, development as a process was always affected from the top down style of functioning. Naturally, because along with our freedom we had inherited a bureaucracy, which was designed by the British to rule, not to serve. The British way of doing things had always been to get things done through a government department and after independence we Indians merely continued this system.

Unfortunately, we forgot that the biggest asset of India is its people. Any sensible government must learn to unleash the energy of its people and get them to perform instead of trying to get a bureaucracy to perform.

What we did at the Kaira Cooperative Union was to give a demonstration of what people's energies can do – what they can acquire and how brilliantly they can perform provided they are guided by honest and sincere leaders like Tribhuvandas; provided their energies are combined with the skills of professional managers to give that energy some direction and thrust. When hundreds of thousands of small milk producers joined together to form cooperatives and when people like Dalaya, myself and many others joined the cooperatives as managers and technical experts, then Sardar Patel's dream of a milk-producers' dairy became a reality. It is not that government officials lack ability; it is that they try to achieve development through a structure that is not designed to achieve it. Our belief at Anand has always been: let the people's energies be unleashed.

~

In 1956 a rather delicate assignment came my way. I visited Switzerland at the invitation of Nestle but with a very specific brief from the Ministry of Industries, Government of India. Industries and Commerce Minister, Manubhai Shah, wanted me to ask the executives at Nestle what they were up to in our country.

Under the excuse of producing condensed milk, they were importing not just milk powder, but also sugar and the tin plate for the cans!

On my arrival at the airport at Nestle's headquarters at Vevey, a Nestle car, about a mile long, was waiting to whisk me off to the best hotel in town where they put me up. I met with Kreeber, one of their two managing directors, and some other officers. The discussions turned pretty heated. I told them that my government had given them a licence to set up a plant in India so that they would produce condensed milk from Indian milk, not from imported ingredients. The Managing Director told me that it was not possible to produce condensed milk from buffalo milk, which was available in India. I said to him, 'If you don't know how to make it, come to me. I will teach you because I believe we can make it out of buffalo milk. I know it is more complicated than making it from cow's milk and there are problems, but they are not insurmountable problems.'

When I assured them that it could be done, they said that their experts would have to come and set up their plant. Then they wanted the entire share capital in their hands. In those days government allowed only 49 per cent share capital to foreigners; 51 per cent had to be Indian. Kreeber said they could not agree to that. So I showed them a way out of that too. I said that 49 per cent could be with Nestle Alimentana and 51 per cent could be owned by Nestle India and in this way the entire project could stay in their hands. I was, in fact, facilitating their entry here.

Ultimately, the Director agreed to set up a plant in India. At this point I told him that they could bring in any number of foreign experts they liked but my government hoped that, in five years, Indians who would be trained for the purpose would replace these experts. Kreeber's response to this was that the production of condensed milk was an extremely delicate procedure and they 'could not leave it to the natives to make'.

At this, I lost my temper. Getting to my feet, I thumped the table loudly and said: 'Please remember that you are speaking to a damned "native". If you are suggesting that even after five years of training, the "natives" are not fit to occupy any position of authority in Nestle you are insulting my country. My country knows how to do without you.' And I stormed out of the meeting – which I hope was what any self-respecting Indian would have done.

On my return to India I narrated the incident to Manubhai Shah. The Minister asked me: 'So then what's the solution?' I said the solution was that Amul would have to make condensed milk.

We got to work and after two years we began manufacturing and marketing our condensed milk. Once this was done, I wrote to the Minister saying: 'Ban the import of condensed milk.' The government issued the ban.

Some weeks later I received a phone call from Manubhai Shah in Delhi. 'Kurien, do you remember this company called Nestle?' he asked. I told him that I did.

'Well,' he said, 'they're here and they want to see me. They've already met my officers and it appears that they now want to set up a plant in India. I remember all the things you told me so I've decided to tell them that I will not see them unless they produce a letter of introduction from you. No doubt they will come to see you and now it is up to you, Kurien, to decide whether Amul can meet all of India's requirements in condensed milk or whether Nestle can be given room to play any role. If you give that letter of introduction I will see them; if you don't give it, I will not see them and they will go away. It is your decision.'

The Nestle team arrived in Anand to see me. There were four executives and among them was Kreeber. As soon as they entered my room, I said, 'Before we sit down to our discussions, I think you should go to our dairy. Look at our condensed milk plant, then we'll talk.' They did as I had suggested. When they entered my room again, I had my opening shot ready. 'Well, Mr Kreeber,' I said, 'what do you think of the "natives" now?'

Kreeber turned red in the face. With the grace to sound humble, he said: 'Sir, I have come here to apologise to you for our rudeness when you visited us. I apologise on behalf of Nestle Alimentana and on my own behalf. Will you please accept our apologies? We want to participate in India's dairy development and we will do it only according to the rules you may lay down.'

Even while I gave the Nestle team the letter of introduction they had come for, which subsequently allowed them to put up their plant in Moga, I knew that once they gained entry into our markets, they would do exactly as they pleased. A multinational never plays by the rules in somebody else's country. This is one of the many reasons why our own economic policies need to be looked at again. It is only logical that nobody will invest money in another country unless they hope to take more money out than they brought in. So how does foreign investment help us? I do not think that foreigners should not be invited in to play a role but I think it should be a role which India decides and which they are required to play in India's interest. Foreign investment can only help us in areas where Indian capital, Indian know-how, is not available.

If the multinationals are desperate to capture the Indian consumer market, which is as large as the whole of Europe put together, then sadly, successive governments have made this easy for them. The only instrument left to us is to defeat them by producing better and cheaper products than they do. In the dairy sector, our cooperatives have continued to hold their own, even against giants like Nestle. If our cooperatives had not been around, we would still be importing baby food, condensed milk and sundry other dairy products just as our neighbouring countries are doing. I take great pride in stating that it is we – our farmers and their cooperatives – who disciplined foreign capital in dairy products in this country.

ON A ROLL

IN A WAY, THE BEGINNING OF THE END OF POLSON CAN BE TRACED EVEN before this period. In the early years of the Kaira Cooperative, when the quarrel between Polson and the cooperative was reaching its peak, Khurody came up with a solution. He suggested to the Minister of Agriculture, Dinkarrao Desai, that the district should be divided into two parts so that one would not collect milk from the other's villages. When I heard about it, I knew we had to be extremely careful about our choice of villages. I decided to draw up a map and take it to the Minister. The map would divide Kaira district into roughly equal parts but the half that generated more milk would be ours, and the other half could go to Polson.

Thus prepared, I went to the Minister. Khurody, Polson and a few others attended the meeting. I presented the proposal to the Minister. I explained that we had divided the district between us and this would put an end to our constant bickering. He looked at my map and said, 'Hmm, seems okay.'

Polson quizzed, 'Where is Naar?'

I explained, 'Naar is in our part of the map.' Polson asked about a couple of other milk-rich villages and I told him those, too, fell in our section.

'Then what are we left with ... ,' Polson began to argue but the Minister cut him short saying, 'we have now discussed it threadbare, there is nothing more to discuss. We have already divided the district.'

As we all walked out, Polson turned around to his contractors' Supervisor – a man who was always armed with a revolver – and asked him: 'Who was that young fellow?'

The Supervisor replied that he did not know.

Polson was furious and shouted at him: 'You don't know? He has walked away with everything and you don't even know who he is?'

But that was not going to be enough. I had to use every opportunity thrown my way and turn it to the advantage of the dairy farmers.

One day we had a VIP visitor to Anand and, after looking around the dairy he came to my house, where I offered him a cup of coffee. As he was leaving he turned to me and asked: 'What did you say your name was?'

'Kurien, Sir,' I replied.

He said, 'I like you. If there is anything you want at any time, call me. And it will be done.' I later came to know that it was T.T. Krishnamachari (popularly known as TTK), the then Minister of Commerce.

Initially, when we started marketing our butter, we found it extremely difficult to put it in the market. The markets were already flooded with the New Zealand Dairy Board's Anchor butter and Pestonjee's Polson butter. I immediately thought of T.T. Krishnamachari and wrote a letter to him:

Dear Shri Krishnamachari,

When you visited Anand, you told me, if I needed anything, you would be able to help. I need your help now. I am not able to market my butter. Will you order a cut on the import of butter by 25 per cent ... .

The reply came back:

> My dear Kurien,
>
> As desired by you I have ordered a cut of 25 per cent … .

There were no meetings, no files, no discussions. After six months, I wrote to him again informing him that our butter production had increased, and asked if he would increase the cut on import of butter to 67.5 per cent.

There was an acute shortage of foreign exchange in those years so my request was received favourably. Once more, TTK wrote back: 'As you desired, I have ordered a cut of 67.5 per cent.'

About three to four months later, I received a letter from TTK, saying:

> My dear Kurien,
>
> As you may be aware, there is a very serious foreign exchange crunch. I have banned the import of butter. Now please ensure that there is no shortage of butter in the Indian market … .

Now how could we fail? It all came together marvellously for us. Usually people ask the government for imports; we were the only ones asking them to reduce it. Welcoming this demand, TTK decided to ban butter import.

Our first product to be branded was butter. Of course, the early years of production were anything but smooth sailing. We encountered some of the strangest of problems.

Polson had always made butter from stale cream. He never collected milk to get fresh cream. The cream merchants would supply him with cans of cream, which went into his butter production. Sometimes these cans of cream would be kept for as long as ten days without refrigeration. Many pollutants – sometimes even maggots – contaminated the cream and turned it malodorous. Polson's Manager, Foster, found an answer to all such problems. He acquired a vacreator – a machine that heats cream for pasteurisation with injected steam that quickly raises its temperature. The machine also creates a vacuum, which removes the steam molecules so that it does not dilute the cream. For Polson, the vacreator served a dual purpose: along with the steam, the vacuum also almost totally removed the foul odour from the stale cream. Some odour though did remain and, ironically enough, became a problem for us at Amul.

Our butter, like butter from New Zealand, was made of fresh cream – milk to cream to butter, all in the same day. When we introduced this butter into the market, people exclaimed in distaste: 'What kind of butter is this? There's no flavour in it. It's flat!' Of course, the Parsis in Bombay city's popular Irani restaurants would not touch it (although I suspect this could as well have been because of their loyalty to 'apro Pestonjee', Polson). This was a serious problem and we had to find a solution quickly. We did. At the end of the butter-making process we began to add a permitted chemical additive called diacetyl, which also gave the butter an added 'flavour'. This solution to a rather unusual problem was legal as long as we printed the line 'permitted flavours added' on the packets. In its new form, Amul butter became more acceptable – and sales showed dramatic improvement.

There were also the niggling but important issues of the amount of salt, moisture and colour to be added. Basic questions had to be debated, discussed and decided upon. Salt is much cheaper than fat so the more salt we added the more money we could make. Besides, the addition of salt gives the product a longer life. But should we cheat our customers? And what should be the moisture content of butter – should it be the maximum permitted limit? One of the basic principles that Amul stood for – giving the customer quality products – would never allow us to cut corners. Value for money was then, and has remained, the fundamental principle for Amul.

Then the colour of the butter had to be considered. The colour of New Zealand butter was naturally yellowish because their cattle always ate green grass, which gives a yellowish tinge to the milk. In Denmark, dairy farmers follow another system – in summer their cattle eat green grass so the milk is naturally yellow; in winter since the cows feed on hay they need to add colour. We used buffalo milk, which is absolutely white and therefore we had to add colour. A number of such delicate balances had to be maintained but, with experience, we could work things out.

Polson was completely convinced that a dairy farmers' cooperative would never be able to make butter. Pestonjee had gone to the extent of getting the manufacturers' nameplates removed from all the machines in his plant, so that nobody could buy those machines and copy his methods. He was an old-timer and his methods of conducting business were old-fashioned. Some of his conventional but impractical ways of working were evident in his office in Bombay. There, he kept a slate behind his desk and one of the daily tasks of one of his staffers was to write down the number of cans of cream that came into Bombay Central from Gujarat and the cans that came into Victoria Terminus from Maharashtra. Once he had the figures Pestonjee would turn around in his chair, cogitate for a while and then decide to keep the price of butter steady or increase it by one anna per seer and cream by half an anna. He conducted this exercise everyday.

However, he was becoming increasingly agitated because Amul butter had been put into the market at a steady price. One day he said to Variava, his Deputy Manager, 'Teach your friend some sense. This is not the way to do business.' He also convinced Variava to persuade me to visit him in Bombay. So on my next visit to Bombay I went to Polson's office. He was extremely hospitable and spent a considerable time talking to me. He then advised me very earnestly to stay with him in his house in Bombay for some time and learn the intricacies of the dairy business. He genuinely believed that prices needed to be changed regularly and that he could help me understand this concept. I did not argue with him. I never wished to offend, much less do anything to harm Polson. The unwritten rule at the Kaira Cooperative was to leave him alone. We simply allowed him to do his dairy business his way and we did ours our way.

Pestonjee always lived in Bombay, which was Polson's headquarters. He owned two dairies – one at Anand and the other in Patna. In those days my house and office were in the same bungalow and every morning at about eleven a.m. Foster, Polson's manager at Anand, would drop by and have a cup of tea with me before going to work. One day he came to my office, he was forty-two years old then, and said, 'You know Kurien, I never thought this Kaira Cooperative would amount to anything or that it would take off the ground at all. But then damn it, you arrived. Once you arrived I knew the future of Polson's was doomed and I've decided to resign from here while I'm still relatively young, so that I can go back home to New Zealand and get another job, start a new life.' And that's what he did. After he left, Variava took over as Polson's Manager.

When we built our dairy, Variava said to Pestonjee, 'You have no idea what this man Kurien has done here. You must go and see the cooperative dairy.' His curiosity aroused, Pestonjee came to visit us. I took him around the plant myself and explained everything to him. He was simply stunned. Then he did something that really touched me. He put both his arms around me and said in all sincerity: 'May Khuda bless you for all you have done here.'

Pestonjee was, indeed, a very fine man. He never did his business in a bad or malicious way. There was always a very healthy competition between Polson and Amul and we had cordial relations throughout. We had agreed that we would never force him out. His business died out naturally when it could not withstand the competition from Amul.

When Pestonjee died in November 1962 his son Minoo took over the management. And the business gradually wound up for good. Unlike Pestonjee, who had started his life with nothing, Minoo was born in the lap of luxury – the type who can turn into a spoiled brat.

Pestonjee knew his son well and left the management of only the Patna dairy to him. The management of Anand dairy went to Pestonjee's son-in-law, Lt Col. Kothawala. One day Minoo came to me and said: 'If you want to ruin anything, ruin the Anand dairy. Don't touch the Patna dairy because that one is mine.' The statement revealed the kind of man he was.

Periodically, Minoo would discuss the sale of the Anand dairy with me. One day he told me that he had spoken to the board and this time he was absolutely serious about selling the dairy. I spoke to our board members, who agreed that we should buy it, and a price was decided. Then Minoo backed out. He came a second time, again offering to sell. Once more I got the board's approval to buy the dairy and again he backed out. When Minoo came to me for the third time wanting to sell the dairy, I ordered him to get out of my room. I told him that if he was serious he should bring his entire board to Anand to meet and talk with our board. He brought his entire board – a very distinguished board – and we discussed the sale and the deal was clinched at Rs 17 lakh.

The next day, Minoo sold the same dairy to a Marwari gentleman for Rs 17 lakh and, some said, took another Rs 17 lakh under the table for himself. The board of directors of Polson were aghast and exceedingly embarrassed. They came to see me and apologised profusely, saying that they never expected he would do something like this. The legitimate amount of Rs 17 lakh went to Polson Ltd, while it is said that the under-the-table amount went into the Devakaran Nangi Trust which later went broke. By some mysterious divine justice, Minoo lost his entire Rs 17 lakh. This was the end of Pestonjee's legendary Polson dairy.

When Minoo sold the dairy to the Marwari gentleman (who bought it only for its real estate value), the first thing the Marwari did was to order the bust of Pestonjee, which graced the entrance, to be removed and thrown out. Variava called up Kothawala to inform him of this and he immediately telephoned me to say: 'Dr Kurien, can you please save my father-in-law's bust from being disgraced?' I promised him that I would and it has since then been given pride of place in NDDB's library, a reminder to all of the role that Pestonjee Edulji played in the history of Indian dairying.

~

The launch of our butter actually began rather inauspiciously. We had already decided to appoint TTK & Sons as our distributors. T.T. Krishnamachari, who by this time was the Finance Minister, owned the company. When the Amul butter advertisements began appearing, TTK was livid as he thought it would compromise his reputation.

He said, 'I give you twenty-four hours. Cancel the appointment.' So literally overnight, we had to find other distributors for Amul. Over time and for many years, Akbarallys were our distributors in Bombay; Spencer and Co in Madras; Empire Stores in Delhi; and James Wright in Calcutta (now Kolkata).

There was no stopping the cooperative once it won the butter war against Polson. Between the late 1950s and the early 1960s, we had established a toehold in the market for not only milk powder and butter but also condensed milk, cheese and baby food.

How we evolved our baby food formula is an interesting story. We were told that the Central Food Technological Research Institute (CFTRI) had already developed and tested a formula. I had requested the National Research Development Corporation (NRDC) to give us a grant of Rs 10,000 which we could give to the CFTRI for the use of their formula.

Dalaya, Molly and I drove down to Ahmedabad for a meeting with Subramaniam, Director of CFTRI. He informed me that Kasturbhai Lalbhai, the Chairman of NRDC, had asked me to meet him at his house for tea. I told them that my colleague Dalaya and my wife would also come with me and I would not come alone. Subramaniam checked with Kasturbhai, who invited all of us. So we all went to Kasturbhai's house. Tea arrived. And then Kasturbhai looked at us and said, 'I expected much older people. I find you are very young. So I better be frank with you.' Then pointing a finger at Subramaniam he said, 'This man is a crook. You think he has made some great discovery. He will give you some half-baked formula and then you will have to spend millions, cleaning up his formula and making it usable. Then you will find that he will go and sell it to Glaxo. You still want that formula?'

I said, 'Yes, Sir. I do.'

'All right then, there is no need to discuss anything further,' said Kasturbhai.

All through that time, Subramaniam was very silent. The meeting was over. All of us finished our tea and got into my car to drive back. Dalaya and Molly sat in the back seat and I drove. Subramaniam sat next to me in the front. He was extremely nervous and kept on fiddling with his tie. He had never expected this to happen. After about half an hour of complete silence he said, 'He (referring to Kasturbhai) spoke very badly ... .'

I said, 'Yes.'

Subramaniam continued: 'He had no business to do so. I don't know what made him say that.'

We returned to Anand and as it happened, Dalaya realised that we could not use their formula at all. He had to change the formula before it could become usable. And yet we had to pay a huge sum as royalty – Rs 30 to 40 lakh – over the years to the CFTRI through NRDC.

Finally, our baby food was ready to be launched. The product was significant for us. We planned to use our milk powder, fortify it with vitamins and set up our baby food in competition with the giant, Glaxo. It was going to be a very difficult goal to achieve because Amul did not have a premium food image – it had an agro, rural image. With our baby food we were bravely trying to enter what, for us, was uncharted territory. We had to make the correct impression on the minds of mothers with new babies without trying to undermine breastfeeding; we had to step into highly hygienic and sanitised areas like nursing homes – and through all this we had to keep in mind that we were challenging Glaxo.

As we had expected, baby food became the centre of controversy. Activist groups came out vociferously opposing the advertising of baby food. However, anticipating this, as part of our advertising campaign, we brought out an Amul Baby Book with a chapter exclusively dealing with the superiority of mother's milk. Some years later we also brought out an advertisement in the press advocating breastfeeding. We took great pains to say that baby food was necessary only in those cases where the mother was unable to breastfeed. And to those who accused us of nearly monopolising the baby-food market, I always pointed out that the monopoly was actually held by the mothers of India, as 93 per cent of our babies were breastfed.

Our methods went beyond merely advertising in the media. The advertising agencies took a holistic view of marketing, beginning with teaching new mothers to reconstitute milk properly so that the babies would not get malnourished. For the first time in India the agency conducted extensive surveys in three cities to find out the behavioural and consumption patterns of consumers. We realised from these survey findings that extensive education was needed. We were the first to see that milk is deficient in Vitamin A and since there is no other source of Vitamin A for the baby, the baby food would have to be fortified with it and also with Vitamin D for the absorption of Vitamin A. To get consumer approval for milk powder that was made from buffalo milk instead of cow milk was a battle too. Certainly it was a constant, uphill struggle but, ultimately, ours clearly became the better baby food in the market.

Interestingly, at one stage, Glaxo approached us, willing to give a manufacturing contract for its baby food to Amul. I said we could consider it on the condition that it would carry the Amul brand name. This so incensed the Glaxo boss that he is said to have declared: 'Amul will never be able to sell its brand of baby food and when their tins begin rotting on the shelves, I will have them collected and thrown into the Arabian Sea!' Such was the arrogance of multinationals.

~

In 1962, the clouds of war against China darkened the nation. I got a call from the Prime Minister's office asking me to come to Delhi for an urgent meeting. Also present at the meeting were some Generals and a senior bureaucrat, Shivaraman. I was informed that the Indian Army needed milk powder, and they asked me how much we could provide and how soon.

I said, 'A thousand tons and within six months.'

One of the Generals looked at me and said, 'That's not enough.'

I said, 'Okay, then 1,500 tons.' They said that, too, was not enough.

'I suppose we stop wasting time, and you tell me the quantity that you need?' I asked.

'We need 2,750 tons,' came the reply.

I asked for a piece of paper as an idea began forming in my mind. I was aware that there was another milk powder plant in

Rajkot, which belonged to the Government of Gujarat. It was a small plant, but I knew that if we put together all the powder and gave it to the army, sacrificing the entire civilian market, then we could fulfil this commitment in six months.

So I did a swift calculation on a piece of paper and said, 'It will be done. Now, can I go?'

The General expressed apprehension, 'Supposing you let us down?'

'That is not the way to speak to Mr Kurien,' Shivaraman said. 'We have the highest regard for his words. If he says that he will do it, he will certainly do it. And besides, may I know what other alternative you have?' That quietened the General.

Then Shivaraman asked me, 'What can the government do for you?'

I said, 'What do you mean?'

He said, 'Loans? Grants? Anything you want?'

I said, 'Mr Shivaraman, you said there is an emergency, and if Amul uses this emergency to squeeze money out of the government, then it is an unworthy organisation. I want nothing.' From that day onwards, Shivaraman was an ally.

On my way back to Anand, I stopped at Ahmedabad, went to the Secretary, Agriculture of the Gujarat government and updated him about the development. 'I have made a commitment, and that commitment is on behalf of the Gujarat government,' I said.

The Secretary assured me that if a commitment had been made the Gujarat government would back me fully. I explained to him that for this to work I would also need control over the Rajkot plant. The Secretary immediately summoned Pheroze Medora, the government officer at Rajkot, and told him, 'Whatever Kurien wants, has to be done. Henceforth, he is the boss. He has made some commitment on behalf of the Gujarat government and we will back him. We will do everything he wants.'

So it was that since we were diverting all our milk and milk powder to the army, our butter disappeared from the market. Minoo Polson decided to take advantage of this and increased the price of his butter. While here we were trying to meet the government's requirement, Minoo Polson was encroaching on our market.

Once more I went to meet the Minister in Ahmedabad. The Minister, who was also from Kaira district, asked me, 'What do you suggest?'

I said, 'Sir, you cannot sit still. You should put an embargo on Polson so that he cannot produce more butter this year than what he produced last year.'

The Minister agreed and an embargo was put. Polson's butter production was frozen. It served the purpose. Minoo Polson was furious and he asked for a meeting with the Minister. The Minister asked me to attend this meeting too. Minoo Polson arrived with his entourage – his lawyer Gagrat, and Variava. They demanded to know how the Minister could do such a thing, how could he freeze their production, by what authority and under what law.

I told them, 'I am giving away 800 tons of my butter market, sacrificing my entire civilian market, and you want to take advantage of it, and increase the price of your butter? How can you do this? This is anti-national.'

The Minister asked them if my accusations were correct and Variava immediately said, 'Yes.'

The strangest thing was that their lawyer, Gagrat, was not aware of this. They had kept him in the dark and he was furious. He asked, 'Is this correct? Why didn't you tell me? You don't have a case to stand on. Come on, let's go.'

They got up and walked to the door where I was standing. When Minoo came near me, I said to him right there in the Minister's office, 'You bloody bastard. You come here, and speak lies to the Minister. I will castrate you.' He turned red in the face. Turning to the Minister he said, 'Sir, see what he is saying. He's saying he will castrate me. Standing in your office, in front of you.'

'You can't control Kurien,' the Minister said to him with a smile. 'You better just go away.'

~

We had one clear advantage in that, unlike private companies impatient to make profits, to sell, to show the shareholders and bosses that they were doing well, we were very patient. It would be a slow and steady progress for us. We kept our sights clearly on the target – the producers must get their due. In our case the producer was a farmer, mostly a marginal farmer. The challenge was to see that the farmers' interests continued to remain paramount. But we knew that even while we kept our farmers firmly in focus, we had to make the products as attractive as possible to the market. We sought help from advertising professionals.

Over time, the accounts for various Amul products were divided among various advertising agencies. While Amul butter by now was the leading brand elsewhere in the country, in Bombay it still lagged behind. In 1966, therefore, the Amul account was given to Advertising and Sales Promotion Company (ASP) with the brief that they should dislodge Polson from its 'premier brand' position in Bombay. This was when Eustace Fernandes of ASP created the Amul mascot – the mischievous, endearing little girl. The image of the Amul girl went down so well with consumers that very soon it became synonymous with Amul. Besides Eustace Fernandes, we can never forget the invaluable contributions and commitment of people like Usha Katrak, K. Kurien and Sylvester DaCunha at ASP. Together, the team at ASP gave Amul butter its memorable and catchy campaign punchline of 'Utterly, butterly, delicious' – which broke all records to become one of the longest-running campaigns in Indian advertising history.

As the number of Amul products began increasing, it was time to go a step further. Till then the system was to bring a lot of empty tins from Bombay to be packaged with Amul products. The quantity of tins kept increasing and we needed several godowns, because fourteen days' stock of tins is a huge number. One day I said to Dalaya, 'Dalaya, this is such nonsense. We bring these damned empty dabbas here. They are not really empty – they have the polluted air of Bombay in them – and they are simply filling up our godowns. Why don't we just use one of those godowns to put up a dabba-making unit?' Everyone agreed that it was a good idea. An entrepreneur, Harshad Kapadia, was already making packaging tins in collaboration with Metal Box. My dislike to partner with foreign companies was well-known. Yet, always quick to recognise the potential in partnerships that could be turned to the advantage of our institution, I persuaded Kapadia to collaborate with Amul dairy. A company called Kaira Can was formed. We gave Kapadia one godown and they put a conveyor belt from there directly to the Amul dairy to transport the tins for packaging. It was a very timely step and it saved us a lot of money.

~

We progressed and evolved step by step. The Kaira Union – or Amul – went on to become a role model for dairy development; we did not know it then, but it was to ultimately give birth to a number of institutions, including India's National Dairy Development Board (NDDB).

Much of the success of the Kaira Cooperative – or the 'Anand pattern' of dairy cooperatives, as it came to be known widely – has its roots in the democratic structure of the experiment. In 1946 Tribhuvandas Patel began this experiment modestly with two cooperative societies and a couple of hundred litres of milk. Today, Anand alone has a district cooperative union with 1,017 village cooperative societies federated into it. Each village has its own cooperative milk society and the farmers become members of their village cooperative. That village cooperative is guided by a managing committee elected by the members. The chairmen of each of these managing committees form the general body of the district cooperative union of Anand. And that district union owns the plant. The general body elects their board of directors to guide and oversee the management of the district union and the dairy plant.

Adhering to the federal system of representation worked wonders in the Kaira Union – system where village cooperatives are represented at the district level through the district federation; and the district unions represented at the state level through the state federations. Democracy – and grass-roots democracy and building democracy through the federal concept – essentially requires that we are accountable to the tier below us. The state is responsible to the district which, in turn, is accountable to the village cooperatives which are accountable to their current 573,962 farmer members in Kaira's 1,017 villages. The quantum of milk collected per day today is 7.39 lakh litres.

Today, the total turnover for this one district cooperative is to the tune of Rs 5,092 million a year, through the sale of its liquid milk and other dairy products. This means that if we deduct manufacturing expenditure, we are distributing Rs 3,943 million to the farmers every year. Each year, because the farmer members produce more, they get back that much more. All this has been happening without anybody giving any subsidy. The farm family is producing a commodity called milk which it markets and on which it makes substantial profits. The dairy cooperative movement, inspired by Amul, is India's largest employment scheme and has more than doubled farm-family incomes.

The story of Amul, however, is not merely the story of garnering profits for the farmers. As the cooperatives started functioning, something far more unique began taking place in the villages of Kaira district. In many ways, the milk collection at the village societies transformed the very social fabric of those communities. Take for example, the farmer and his wife. The farmer has a couple of acres of land, sometimes less, which he looks after and his wife looks after their one or two buffaloes. She invests a lot of time and energy in caring for the cattle and therefore, for all practical purposes, those are her animals. With the dairy business picking up in Kaira, her income through milk soon became equal to her husband's income through his land and that led to a different kind of equation between them.

There were other implications too. From the very beginning the village cooperative societies insisted that the queue for milk collection would be formed on a strictly first come first serve basis. Imagine the situation when 5,40,000 farmer members stand in line at 960 village milk collection centres across the district, irrespective of sex, religion and caste. What does it do to a 'high-caste' Brahmin to stand behind a Harijan because he came after him? What does it do to the Harijan to stand in front of the Brahmin because he came earlier? Is it only an orderly milk collection? Is it not a blow to the caste system? What does it do to both of them to watch their milk flow into the same can? In another situation, a villager whose buffalo has run dry buys the milk from the centre that has been collected from both the Harijan and the Brahmin. Do these experiences not break down caste prejudices? Of course, we were never so naïve as to believe that the caste system would be eradicated overnight – it is, unfortunately, much too strongly and deeply embedded in the Indian system. But what the process certainly did was to begin chipping away at such abhorrent biases, and some new practices – such as a queue system regardless of caste hierarchy – were accepted because of the milk collection and payment system which we put in place through the cooperatives in Anand.

Our villages in India are dirty; our village milk collection centres were also somewhat dirty, but they were distinctly cleaner than the rest of the village because somebody sprinkled water to keep the dust down, somebody put some glazed tiles outside to keep it clean, somebody used a flit-gun to keep the flies away, someone used sanitizers to wash the pots and pans. These were the first few steps towards sanitation. How could we talk of clean milk production unless we talked of sanitation?

Very early in the cooperative's history, we realised that a team of dedicated veterinarians was vital to the project. Soon, one of Kaira Union's biggest strengths was this group of veterinarians, who visited every village once in fifteen days and treated all sick animals without charge. Take the case of dystokia, where a pregnant buffalo cannot calve because the uterus is twisted. Every villager who owned a buffalo knew that this was almost always fatal for the animal. There was a time in our villages when all that the villagers could do was to commiserate with the woman who owned the buffalo. Now, through telephones, they send for the cooperative's veterinarian – their veterinarian. Day or night, Monday or Sunday, the veterinarian gets there within hours of being called. He sees the problem immediately and he rolls up his sleeve and gets to work. He lies behind the buffalo, puts his hand into the vagina, he rolls one way and has the buffalo roll the other way and he successfully brings the calf out.

This efficiency shown by our veterinarians led the villagers to question the inefficiency of medical facilities provided to them. Therefore the Cooperative Union thought of beginning healthcare programmes for the villagers, especially for women and children, and the Tribhuvandas Foundation was set up.

Once the Amul dairy was ready we initiated a programme whereby we took village women – members of our cooperative societies – on guided tours to show them their dairy. They were proud to enter this modern Amul dairy, to see the lush green lawns, the magnificent buildings, the gleaming stainless steel and to know that they owned a part of it and had a share in its profits.

Similarly we began taking them to visit the cattle-feed plants and explaining to them how feed is formulated, why it has 18 per cent protein, and what protein is. We informed them why we add vitamins and minerals; why a female buffalo who is dry in the last months of pregnancy should be fed properly even though she is not giving milk, as a foetus was growing inside her. Could these women not relate all this to their own growing foetuses? Were we only discussing animal nutrition? We started taking our women cooperative society members to the artificial insemination centres, too. Their only demand was that no male member from their village be present during these visits. We showed them how semen is collected from a bull; we made them look through microscopes at live semen; we explained to them the mysteries of conception and birth by using charts. Did they then not ask: 'Is this not what happens in humans beings too?' Could we not then lead them on to birth control?

What then was the Kaira Cooperative? It was certainly not only about milk. It was very soon becoming an instrument of social and economic change in our rural system. It was evolving into a programme that involved our farmers in their own development. This I learnt very early on through my years of working closely with Tribhuvandas and the farmers of Kaira district: true development is not development of a cow or buffalo but development of women and men. However, you cannot develop women and men until and unless you place the instruments of development in their hands, involve them in the process of such development and create structures that they themselves can command. What, therefore, is a government at its best? It is a government that 'governs' least and instead finds ways to mobilise the energies of our people.

Amul – or the Kaira Cooperative, or the 'Anand pattern' of

cooperatives – demonstrated without a doubt that the biggest asset of India is its people. Over five decades ago I was compelled to come to Anand. What began as 'bonded servitude' turned out to be five decades of reward and satisfaction. The only credit I can claim is that I had the good sense to remain in Anand in spite of opportunities that came my way to go to Delhi or other places. I chose to remain in Anand, as an employee of farmers, all my life. It was the wisest decision I have ever taken.

During the 1960s I came across the writings of Barbara Ward, a famous economist. She had a remarkable career which included advising two US Presidents, John F. Kennedy and Lyndon B. Johnson. It was John Kennedy, I recall, who said that one of the pleasures of the world was to sit at her feet and listen to her talk. I was fortunate to have the pleasure of listening to her when she came to Anand. She was an extremely beautiful and inspiring person in many ways.

Ward, perhaps more than anyone else at that time, helped to map out an international agenda for alleviating poverty and tackling its causes. She urged the world to work towards the sane and sustainable management of our planet's resources. She believed that 'with equal social and economic emancipation', poverty can be tackled within the limits imposed by ecology and the availability of resources. She also felt that there is but one constraint preventing a solution to the twin problems of poverty and environmental degradation – a constraint which could well prove insurmountable – and that is, 'whether the rich and fortunate are imaginative enough, and the resentful and underprivileged poor patient enough, to begin to establish a true foundation for better sharing, fuller cooperation and joint planetary work'.

It was a concept that deeply influenced me and I pledged to myself that, through my work with Anand's cooperatives, I would try to make a modest contribution to the vision of people like Sardar

Patel and Barbara Ward. To this extent I believed myself to be a part of a mission that would prove that the ultimate end of developmental programmes is not merely the mechanical modernisation of the production of milk, but rather, it is the beginning of a process where our rural majority can control their future and build for themselves a richer and more satisfying community.

When I started my working life as an apprentice with the Tata Iron and Steel Company, J. R. D. Tata was not running the steel plant; his American General Manager headed it. Tata productively combined the professional management he hired, with his money power, common sense, ability and character. But without that professional management the company may never have succeeded. Today India is among the most industrialised nations in the world. How did this happen? It happened because we combined our native shrewdness and the money our great industrialists had with professional management. I was convinced that the biggest power in India is the power of its people – the power of millions of farmers and their families. What if we mobilised them, if we combined this farmer power with professional management? What could they not achieve? What could India not become?

But before this could happen, that 'true foundation for better sharing, fuller cooperation' which Barbara Ward spoke about had to be built. If the stark imbalance between the cities and villages, between industry and agriculture in our country was to be corrected, it became necessary that farmers be organised. Keeping this in mind, it became necessary for us at Anand to erect structures that would include our people into the decision-making process. What use is democracy in Delhi if we do not have democratic institutions at the grass-roots level? What is a village society if not a school where our future leaders should learn how to manage their own affairs? What is a district union if not a college where the elected representatives can get postgraduate training in management and business?

There is nothing wrong in building flyovers in Delhi. What is not fair is when we do not also build an approach road to villages across the nation. There is nothing wrong in having fountains with coloured lights in the capital. After all, Delhi should be beautiful. But it is unjustified when we have not provided drinking water to all our villages. There is nothing wrong in having a modern, private hospital in Bombay, or the All India Institute of Medical Sciences in Delhi, or other large medical institutions in our big cities. But it is not justified when we have not arranged to have two drops of a medicine put into the eyes of a farmer's newborn baby, and that baby goes blind. While this would have cost us nothing, we have preferred to spend crores of rupees in building five-star hospitals in cities. Why does this happen? Because policy making is in our hands – in the hands of the elite – and naturally, even unconsciously perhaps, when we make policies we make policies that suit us; we usurp the resources of this land somewhat shamelessly to benefit ourselves. The most charitable interpretation of it is that we do it unconsciously.

I opted to remain an employee of farmers all my life, not because I could not get a job in the city of Bombay or any other city anywhere else. It was only because I felt that I had the best job that I could ever get. Nor did I do it out of any great nobility of character – I did it because I realised I had a job which gave me the greatest pleasure, the greatest satisfaction. The idea of working for a large number of farmers translated itself into the concept of working for social good. I soon realised that money is not the only satisfaction that one can seek, that there are several other forms of satisfaction and all of these were available to me at Anand.

~

Soon, Amul's success began making waves across the nation. Word of the miracle of this cooperative dairy tucked away in a small corner of Gujarat spread far and wide. Anand emerged out of its earlier anonymity to find a place on the country's map. People flocked to this small town to see how this 'Anand pattern' actually worked. Was it really possible to bring welfare to the people without the help of the government? Could such an organisation truly function without the expertise of bureaucrats? There were those – even among the government – who acknowledged my role as a professional manager who was notoriously ruthless in bringing to the farmers what was rightfully theirs. One such person was C. Subramaniam, the Union Minister for Agriculture in 1964.

One day Subramaniam summoned me for a meeting in Delhi. The meeting was scheduled for three p.m. that afternoon. I had never met C. Subramaniam before and had no idea what to expect. I got there five minutes before the appointment and was asked to meet the Joint Secretary who seemed very tense and kept looking at his watch nervously. I asked him what the Minister wanted from me but he said he had no idea. At exactly two minutes to three, he jumped up from his seat, took me down the corridor to the Minister's room and stopped in front of the door. He looked at me. Very politely, I said, 'After you,' and waited. He shook his head vehemently, saying: 'I only came to show you the door. I am not supposed to be in there with you.' Such was the awe surrounding C. Subramaniam.

I entered the room alone and saw this slightly built man, dressed in sparkling white clothes, sitting behind a huge desk. I introduced myself. He asked me to sit and came immediately to the point: 'I want you to take over as Chairman of the Delhi Milk Scheme – DMS – under such terms and conditions as you may stipulate.'

I was taken aback, to say the least. I told him that I was not looking for a job and I did not have any intention of coming to Delhi.

'Did you not hear me?' the Minister asked sternly. 'I said on such terms and conditions as you may stipulate. So stipulate your conditions.'

I told him my main condition – that I would not leave Anand. He looked somewhat dismayed and said, 'Then what shall I do? I was told that you are the only man who can set things right at DMS. So tell me, what can you do for me?'

I told him I would take a look at DMS and tell him what to do with it. 'Okay,' he said. 'The government accepts your proposal.'

When I started scrutinising the working of DMS it was like opening a can of worms. There was absolutely nothing that was being done right. Indeed, for us it was an unforgettable lesson in what ought not to be done. First of all, DMS followed the bizarre and irrational practice of rationing milk. Since milk was a rationed commodity, if you were living in Delhi the authorities obliged you by giving you a milk card. In addition and beyond all reasoning, milk was being sold at far below the cost price. I was aghast. Why should the people of Delhi get milk at subsidised rates when the rest of India paid market rates? Those days, people who were given the milk ration card in Delhi were proud to have the privilege of holding this card because it would get them DMS milk at Rs 2 a litre when the market price was Rs 5 a litre. Of course, the system was grossly misused, as such systems invariably are. There used to be serpentine queues at the DMS booths each day. Total chaos reigned early in the mornings outside these booths since goons would be waiting to corner as much of the supply as they could and sell the milk at a premium.

Amongst a host of other harmful practices, I noticed that all the milk was supplied to DMS by milk contractors who procured it from the farmers. This was a recipe for disaster since the contractors' lobby was an extremely powerful one in Delhi, one which controlled many politicians as well as the DMS management. So powerful was the contractors' influence that no one at DMS had the guts to reject even substandard milk. DMS had its chilling centres at different places and the contractors supplied milk to these centres. I felt that these chilling centres should have been managed by the states in which they were located and DMS should be buying milk from them rather than trying to manage the centres itself.

DMS was designed and established by the government to supply liquid milk to Delhi's consumers. However, they never seemed to collect enough milk to supply to the city's inhabitants and had to resort to importing milk powder in order to reconstitute it into liquid milk. It was perplexing, therefore, to find that even when they did not have adequate milk to supply to the people, they had set up a milk powder plant. The rationale for a milk powder plant can only be surplus milk. What is more, DMS had also set up plants for making flavoured milk, curds, butter, ghee and other products. It became only too obvious to me that all these had sprung up to provide Delhi's pampered politicians and bureaucrats with dairy products at subsidised rates. In turn, the government was subsidising the dairy from taxpayers' money. It was absolutely shameful.

After a couple of days of intensive investigation into the working of DMS, I prepared a report and went back to C. Subramaniam. I said to him, 'Sir, I must inform you that things are very bad. Everything is in a terrible mess. How can a government ministry hope to run a dairy industry?'

'Well, how long will it take you to clean up the mess?' he asked.

'I can clean it up, Sir, but the question arises, what are my powers?' I asked.

He looked at me unblinkingly and said, 'Your powers will be those of the Finance Minister and the Agriculture Minister.'

I was stunned. Never before had anyone been given the powers of two Union Ministers. 'Sir, with due respect, you can lend me your powers but how can you give me the powers of the Finance Minister?'

'I didn't know you at all,' he explained. 'It was Finance Minister, T.T. Krishnamachari, who told me: "Send for Kurien if you want to clean up this place. Put him in charge – he's the only one who can do it." That's why I've called you. So you see there's no problem in giving you the powers of these two ministers. Now tell me, how long will it take?'

I told him it would take me six weeks to clean up DMS. He was very surprised and looked extremely sceptical. I also told him that I knew only some aspects of the operation but there were others who knew the other facets and I would need to have a team working with me if we were to overhaul DMS. 'The government accepts your suggestion,' he said in his inimitable style. 'Build your team and let me know who they are within seven days.'

And then, to my surprise, he added: 'I am sacking Dr Sikka, the current Chairman of DMS.'

L.C. Sikka was a very able man. He was highly qualified and he knew his job well. 'But Dr Sikka is a senior colleague of mine in the dairy industry,' I said to C. Subramaniam. 'Must you sack him, Sir? Can you not give him an honourable exit?'

'We'll see,' said the minister cryptically.

I asked him if I could go and meet Sikka and tell him what was happening. He allowed me to do so and so I called Sikka and asked him if I could see him. Sikka did not have a clue about the developments, and so I briefed him about my meeting.

'What the hell does he mean?' Sikka exclaimed. 'I've been appointed by the Public Service Commission. How can he sack me?' He was understandably upset.

'I don't know the rules, Dr Sikka,' I said to him. 'But I do know this much – this Subramaniam is a ferocious man. I wouldn't like to tangle with him and I would advise you also not to mess with him. My advice to you would be that you should quit before he sacks you.'

Just then the phone rang. It was the Minister's Special Assistant, R. Venkataramanan, who later became the Governor of the Reserve Bank of India, informing Sikka that the Minister would like to see him. Sikka looked at me and said, 'So it's true?'

I said to him, 'Dr Sikka, fond as I am of playing practical jokes, would I come to you and do something like this?'

'Then would you like to come and inspect DMS?' he asked.

I assured him that I was not there to inspect any dairy but that I had come to him because he was a colleague and I wanted to give him the information that I had.

C. Subramaniam called Dr Sikka to his room and asked him to submit his resignation. That is how I got involved in the cleaning-up of DMS. There was so much corruption. So many vested interests on whose toes we had to step on. But we cleaned it up. I hand-picked my core team of seven people, which included Dalaya and Michael Halse (of Food and Agriculture Organization) from Anand and N.S. Dave, the head of milk distribution from the Bombay Milk Scheme. People were surprised at my choice of Dave because Amul and BMS were seen as rivals, but I knew without a doubt that Dave was the best man in milk distribution. For the duration of six weeks, Dalaya was stationed in Delhi and I divided my working week between Anand and Delhi. We worked round the clock for forty-two days. We played merry hell with the DMS – we had to. Within forty-two days we mopped up the mess and it became a brand new dairy.

How did we do it? We brought in another ninety-two people – experts in their respective areas – from Amul, from the cooperatives, who actually worked in the dairy. These ninety-two people were not visible but they worked tirelessly in the dairy. Even the ministry did not know of their existence. In Delhi they saw only the seven of us from the core team.

Most of the employees at DMS were on deputation and we began the clean-up process by reverting all those who were useless. The way the organisation worked, it came as no surprise that it had been running at a huge loss for years together – something to the tune of Rs 10-12 crore per year. I told the government that if they were serious about changing things, first and foremost, they would have to stop all the subsidies. After all, who were they subsidising it for? They were subsidising for the privileged class and favouring the contractors who were exploiting the farmers. A shock wave went through the government and it caused tremendous turmoil. It shook up the politicians and vested interests in Delhi. Till then, the milk price in Delhi was being fixed by the Central Cabinet. I got that changed as well. Removing rationing meant that I first fixed the price of milk to what I thought it should be so that the dairy did not lose any money on it. The Minister, too, was worried about what would happen when we removed the rationing system, but he had full confidence in me.

As part of the overhaul, I also decided to get the DMS milk booths redesigned. The older ones looked dismal. I wanted something that looked friendly and I also wanted the Delhi consumer to know that changes were taking place for the better. My team worked hard behind the scenes, streamlining the system and getting the milk procurement process in perfect order. Once everything was ready, we scrapped rationing overnight at DMS. Suddenly, milk was available for everyone in Delhi at the same price. C. Subramaniam was relieved when he saw that nothing untoward happened.

A lot of people may call the way I conducted the DMS operation reckless. But I was not reckless. I knew that changes like these must be effected overnight – they should not be brought about gradually, giving the vested interests ample time to mobilise and create trouble. Of course, it could have misfired. However, I was game to take whatever consequences came my way. As it happened, luck was on our side. About the only reaction that took place was that some twenty women – supposedly housewives –

were brought to Krishi Bhavan the next day to agitate. The media, which by now was clearly on our side, reported: 'Twenty healthy, obviously not starving women, went to Krishi Bhavan to protest about the removal of rationing ... .' And at the end of our six weeks, when we went to DMS to say goodbye, we were gheraoed. The ministry had to send armed police to bring the situation under control.

I know that by the end of this entire process, I made a lot of enemies in Delhi's corridors of power. It did not bother me, however, because during this period I also found a friend in C. Subramaniam. We hit it off extremely well and he supported me throughout. He saw the way I worked and understood the values that I brought to my work and I think he grew very fond of me. In turn, I became his lifelong admirer. It had taken a person of his stature, integrity and strength to take the first step to set things right in at least one area of government's functioning.

Once our six weeks were over and my team had delivered a changed, clean DMS, the Minister wanted a new chairman. He thought that we should put some senior defence officer in charge and that way perhaps some discipline could be instilled into the organisation. We interviewed some defence people but could not find anyone appropriate among them, and so we looked at other groups and finally we came down to IAS officers. After a long unsuccessful hunt, Subramaniam asked me to find someone, saying that he was fed up of looking for a competent chairman himself. I told him that the only IAS officers I knew were in Gujarat. He assured me that it did not matter if I got someone from there. This post was valuable not in itself but it was godsend for bureaucrats who wished to get to Delhi. I suggested the name of H.T. Sadhwani, one of the secretaries from Gujarat. He was a good man, mild-mannered, gentlemanly, highly recommended by the Home Secretary. I thought that at that juncture DMS needed a gentleman in charge. When C. Subramaniam was finally taken to inspect the cleaned-up DMS, it was the new Chairman, Sadhwani, who showed him around.

My difficult assignment successfully completed, I turned my full attention once more to Anand, plunging myself again into the ever-increasing work of the cooperative.

~

In 1957 the Kaira Union registered the 'Amul' brand. While that year was a momentous one for us at the cooperative, it had been an equally propitious one for me personally. In 1957, Molly and I were blessed with a baby daughter. We named her Nirmala. With her arrival our little family was complete and her innocence and purity filled our home with unbound joy.

The years had passed quickly and by the time we finished the DMS assignment, Nirmala was already a lively and loving seven-year-old. One day, none other than Maniben very starkly brought home to me the fact that I was not spending enough time with my daughter because of my work. When I got home that evening, it was rather late. Maniben, who was on one of her regular visits to our home, was sitting in the verandah, spinning khadi. As I walked in, she looked at me sternly and said, 'So you have come? Now sit down here.'

I obeyed her bidding and then she said to me, 'Do you know that I was known as one of "the three witches of Delhi"?'

'Who were the other two, Maniben?' I asked.

'Rajkumari Amrit Kaur and Fatima Jinnah,' she replied. 'And do you know how I became a witch?'

'No, Maniben,' I said, by now quite intrigued.

'My father was always very busy fighting for the independence of India,' Maniben began explaining. 'He had no time for me. He was always busy and in our traditional Gujarati families, the fathers cannot even hug or cuddle their daughters. My mother died very early and although my father loved me very much, he could never show it. So I was brought up without any love and I never got married. When you are brought up without any love like I had been, you become a witch. Do you want your daughter to become a witch?'

'No, Maniben,' I said in some horror.

'In that case, get home by six o'clock,' she ordered me. 'Nirmala has been waiting here for you for a long time. Eight hours for your dairy, eight hours for your family and eight hours for sleep. If you don't follow this, your daughter is bound to become a witch.'

I took her advice to heart at that moment but I confess, despite my good intentions, I could not spend as much time with Nirmala as Maniben had prescribed. Most of the time it seemed as if my entire life was consumed by work and I was conscious that Molly had to constantly step in to meet most of the demands for time and attention from our growing daughter. I took succour in the fact that we had some extremely close and loving friends who treated my family like their own.

Once while returning home from school, Nirmala got into a wrong bus, got lost and landed up in a nearby village. Such was the sense of familiarity and trust, that all that this little girl did spontaneously was to go to the milk cooperative society in that village and the people there brought her straight home to Anand. All of Anand had turned into a supportive, extended family. Unfortunately, as there were no high schools locally we finally sent Nirmala to St Mary's school – a boarding school in Poona. We missed her sorely.

Besides Maniben, Morarji Desai, too, who was very fond of us, was a regular house guest. Whenever he visited Anand he stayed with us. Both Maniben and Morarjibhai were strict vegetarians but they always insisted on eating their specially prepared meals with us, at the same table where often Molly and I ate non-vegetarian food.

During this period, Jawaharlal Nehru visited Anand twice,

once accompanied by his daughter, and each time they, too, stayed with us. It was during one of these visits Dalaya and I had one of our rare quarrels. An official function had been held at the dairy and a huge crowd had turned up to see the Prime Minister. After the function was over, the staff was keen that the Prime Minister should walk through the crowd. Just as he was preparing to do so, one of his security officers from behind shouted, 'The Prime Minister will not walk through the crowd.' And like an obedient schoolboy, he turned around and came back and we returned to my house.

At the dairy, the people waiting to see him were, naturally, very disappointed. They turned restive and said that they wanted to see the dairy. Dalaya, who was there, ordered the workers at the dairy not to allow the crowd inside. Soon I got a call from the dairy telling me that the farmers were getting upset and were making a scene. They were saying that it was their dairy and they would like to see it. I told the people at the dairy, 'Let them in.' Dalaya came to me in panic and said, 'What are you doing? They will destroy the dairy.'

I said, 'Dalaya, let them destroy the dairy. We can always rebuild it. But if we destroy the institution, we can never rebuild that. Let them in.'

Dalaya was furious and he strode off saying that he was going to quit. Morarjibhai came to know of this and he went to Dalaya's house and calmed him down. The dairy remained unharmed even after the crowds had seen it, and Dalaya continued with us.

A BILLION-LITRE IDEA

IN 1964 THE KAIRA UNION'S NEW CATTLE-FEED COMPOUNDING FACTORY sponsored by Oxfam was ready at Kanjari, approximately eight kilometres from Anand. It was a modern, automated plant, a first of its kind in India. This was a revolutionary step for the dairy industry of the country. We thought that such a plant should be inaugurated by the nation's Prime Minister. We invited the Prime Minister, Lal Bahadur Shastri, to come to Anand and officially commission the plant. The occasion was to be Sardar Patel's birth anniversary, 31 October. Shastriji accepted our invitation.

Anand, however, did not have a hotel fit to accommodate him. The accepted practice was that I would host the VVIPs. It was sometimes a tiresome business because of the security drill. For instance, once a state governor on his way back from Ahmedabad to Baroda wanted to have breakfast with me. His security personnel turned up to inspect the facilities at my house even before the household was awake. They measured the bed in the room where the governor was to rest and, to their horror, found it short of the regulated length (the governor was an unusually tall man). Getting a bit weary of all the nit-picking, I told them that His Excellency would just have to lie diagonally.

Shastriji's demand proved to be even more complex. He made an unusual request to modify the programme we had prepared for him, sending many of us into a bit of a tailspin. He sent word to the Chief Minister of Gujarat, Balwantrai Mehta, that he would like to come a day earlier and spend a night in a village as the guest of a farmer – preferably a small farmer in Kaira district.

As far as I knew, no Indian prime minister had ever asked to stay in a village, so naturally this unusual request caused some consternation. The Chief Minister asked me to help them arrange this stay. I told him that if India's Prime Minister went to a village, at least three hundred policemen would be dispatched to that village even before he arrived. Most villages in Kaira district had an average population of around three hundred and with such a strong concentration of policemen the village would resemble a police camp. Why would the Prime Minister want to go to a village to see a police camp? However, if the Prime Minister really wanted to see a village in its normal and natural condition, the Chief Minister would have to entrust the Prime Minister's security to me.

Balwantrai Mehta sent for F. J. Heredia, the Home Secretary, Gujarat, and informed him of my suggestion. Heredia was not at all convinced. 'This will not do,' he said. 'If something goes wrong, it's my neck on the line and not Kurien's. I'm sorry I cannot agree to this. The security of the Prime Minister is my business and I will not delegate it to anyone.' But he did understand the point I was trying to make and since we were friends he promised to arrange it in a way that would meet his needs as well as mine.

'How will you manage it when Kurien insists that there should be no policemen and you say there have to be policemen?' the Chief Minister asked him.

'It's quite simple,' explained the Home Secretary. 'No one – simply no one – should know that the Prime Minister is going to the village or which village he is visiting. Then the Prime Minister would be safe.'

This seemed to make sense. Secrecy was to be the basis of our security arrangements.

'In that case, you and Kurien arrange everything,' Balwantrai Mehta agreed.

Heredia and I met. We picked Ajarpura, a village a few kilometres from Anand. Ajarpura had one of the oldest registered milk cooperatives in the district. I also identified the farmer – Ramanbhai Punjabhai Patel – and explained to him that two foreigners were visiting us; since they wished to spend a night at the village, could he arrange for their stay? Ramanbhai was perplexed as to why any foreigner would want to do that. I convinced him that these foreigners were a bit quirky and asked him if it would matter if they stayed in his house for one night. I asked him not to do anything special except tidy up a bit and clean up the bathroom. He agreed.

On the day of the Prime Minister's visit, at about five-thirty p.m. the guard of honour was kept ready and all the official arrangements were made to receive him at my house in Anand. The ministers, too, had arrived. At this stage, I called the Collector and handed him a sealed envelope.

'What's this?' the Collector asked in surprise.

'It's merely a letter signed by the Home Secretary which says that there's a slight change in the Prime Minister's programme and you will now take instructions from me,' I told him.

The Collector and I then drove to the village. Ramanbhai, after sprucing up his hut and sprinkling water to keep the dust down, was waiting for his two 'foreign guests' to arrive. I went to him and said, 'Now you should know who your two guests really are. They are the Prime Minister of India and the Chief Minister of the state of Gujarat.' 'The Pradhan Mantri in my house? What have you done to me, Saheb?' exclaimed Ramanbhai in anguish.

'Nothing,' I said, trying to calm him. 'Believe me, they're good people. Just as good as you and I. You treat them as you would treat any guest of yours.'

'Saheb,' he said, 'I have not cooked anything special. You told me not to.'

I assured him that they did not want anything special. I introduced him to the Collector, the head of his district, and then I said to the two of them, 'I leave the Prime Minister to both of you. You look after him now and I'm going home.'

I explained to them that Shastriji had no fixed programme. He would come here and decide what he wanted to do while he was a guest of the village. I told them that I had to return home because my wife was there, coping alone with all the other guests who had no idea that the Prime Minister would not be arriving that day.

According to the plan, as the Prime Minister's convoy drove from Ahmedabad to Anand, the Prime Minister's car alone was diverted to Ajarpura village while the rest of the convoy proceeded to Anand.

Shastriji reached the village in time, met Ramanbhai and his family and shared their simple dinner. Then he went for a walk around the village. Even though he was recognised, he moved around freely among the villagers. He invited himself into their homes, sat with them and their families and talked to them at length. He wanted to know about their lives, what the women of the village did, whether they owned any buffaloes, how much milk they produced, how much were they paid, did they get any incentive to increase production, why were they members of this cooperative and how their village society was faring. He continued the stream of questions at great length.

He visited the huts of Harijans in the village. He sat with them and talked to them. He visited the Muslim families in the village. Till two o'clock in the morning, he was busy talking to the farmers and their families about their lives and their problems. The Home Secretary had to remind him about his next day's programme, which was to begin at seven a.m. He was forced to retire for the night.

The next morning the Prime Minister visited the village milk cooperative society run by the elected representatives of the village. I met him there for the first time and explained to him the working of the cooperative. Only after this did he come to Anand and to my house. Later, he declared open the cattle-feed compounding factory and addressed the gathering with an inspiring speech. Then we returned to my house.

At home, he sat me down and told me something extremely interesting. He said, 'Under the Second and Third Five Year Plans, we have built so many dairies. All of them owned and run by the government. All of them unmitigated disasters, running at a loss. But I heard Amul dairy and its products are liked throughout the country. It is available throughout the country and has an extremely high growth rate every year. I want to know why this particular dairy is a success when all the others have failed. That is why I decided that I would stay here and find out. And that is why I spent a night with the villagers, trying to fathom the reasons for the success of Anand's Amul dairy. But I am sorry to say, Kurien, that I have failed.

'I looked at the soil. Good soil, but not as good as the Indo-Gangetic plains. I asked about the climate here. Cold in winter, very hot in summer, I was told. So it is in most of India. Nothing special. I enquired about the rainfall. Thirty inches of rain for three months of the year during the monsoon – much like the rest of the country. I had expected to see the entire landscape green, with cattle grazing contentedly, but the whole place is brown, just like the rest of India. I did not find any abundant availability of fodder and feed here. I looked at your buffaloes and don't mind my saying this, Kurien, but they are not as good as the buffaloes in my home state of Uttar Pradesh. Those buffaloes are certainly better and even give more milk. Lastly, I looked at your farmers. They're good people – farmers are always good people – but they are not as hardworking as the farmers of Punjab. I can't find a single reason why Anand is such a great success. Now, can you please tell me what is the secret of its success?'

I assured the Prime Minister that all his observations were absolutely correct but that there was one difference, which he had failed to notice. The solitary difference was that Amul dairy was owned by the farmers themselves. The elected representatives from among the farmers managed it. These elected representatives had employed me as a professional manager to run their dairy. I was an employee of the farmers.

In this dairy that was owned by the farmers, therefore, my job as a Manager was to satisfy the farmers who supplied milk to the dairy. I had to provide the infrastructure to the farmers to help them increase production. I had to ensure increased production so that they benefited. I could never refuse to collect the milk they supplied. This was a dairy that was sensitive to the needs of farmers and responsive to their demands. I explained to the Prime Minister that just as in Anand, in all advanced dairying countries, the dairies were owned by farmers. I pointed out to Shastriji that all we had done at Anand was to prove that what was true for New Zealand, Denmark, Holland and even the US, was also true for India.

They employed me, a professional who, in their judgement, was capable and honest. They were satisfied with my trustworthiness, competence and honesty. They left me free to run the cooperative as I thought best. What is more, they had protected and supported me during the initial stages until I found my feet and did not allow anyone to interfere with my work.

The Cooperative Societies Act of India is a stagnant act. It does not encourage the creation of truly democratic institutions. It is nothing but an appendage of the cooperative department of the government. But the Kaira Cooperative – Amul – in spite of such an act was a true and functioning cooperative because of the efforts of its Chairman, Tribhuvandas Patel, who was selected by

Sardar Patel, and the farmers had complete trust in him. I explained all this, at great length, to our Prime Minister.

The Prime Minister, who had been listening to me avidly, looked excited and said, 'Kurien, this means that we can have many Anands. There are no special reasons to have an Anand only in Gujarat.'

I nodded my head in agreement.

'So then, Kurien,' he continued, 'from tomorrow you shall make it your business to work not just for Anand, not just for Gujarat, but for the whole of India. The Government of India will give you a blank cheque, it will create any body, any structure you want, provided you will head it. Please replicate Anand throughout India. Make that your mission and whatever you need for it, the government will provide.'

I heard him out and then told him that before I could agree to his request, I had certain conditions. The first was that I would remain an employee of farmers. I would not be an employee of the government. I would not accept a single paisa from the government. When Shastriji wanted to know the reason for the condition I told him that an employee of the government inevitably has to please his superiors; an employee of farmers has to please only the farmers.

My second condition was that the new body, responsible for replicating Anand throughout the country, should not be located in Delhi. 'People in Delhi think about many things but they hardly ever think about farmers,' I reasoned. 'In Anand we can think of nothing else other than farmers, agriculture and dairying. We have no other interests. So whatever body the government creates must be located at Anand. I refuse to move to Delhi.'

The Prime Minister agreed to both these conditions.

~

Prime Minister Lal Bahadur Shastri was a people's leader in every sense of the term. He came to Anand, saw the working of the cooperatives and recognised them as instruments for social and economic change in the rural sector. If there was anything that was closest to the heart of this diminutive but great man, it was the well-being of our country's rural people.

Shastriji left Anand and headed back to the capital. In December 1964 he sent out a DO (demi-official) letter to his cabinet ministers and to the chief ministers and governors of all states, in which he stated his intention of launching a programme which would form Kaira Union-type cooperatives all over the country. 'We envisage a large programme of setting up cooperative dairies during the Fourth Plan and this will no doubt be based on the Anand mould ... ,' the letter stated in no uncertain terms.

Soon enough I was invited to Delhi for a meeting with the Minister for Food and Agriculture, C. Subramaniam. I had already worked closely with him when he had asked me to overhaul the Delhi Milk Scheme. At our meeting, the Minister discussed with me the idea that the Prime Minister had mooted regarding replicating milk cooperatives in the country and then he asked me how much money I would need to start off the project. Bearing in mind that we would need to buy some basic office equipment such as typewriters, tables and chairs, I told him that I would need Rs 30,000.

Subramaniam laughed. 'Rs 30,000?' he asked in amazement. 'What are you going to get with Rs 30,000? I thought you'd ask for Rs 30 crore – even Rs 300 crore ... .'

I assured him that Rs 30,000 would be sufficient and he promised me he would arrange for the funds. But his bureaucracy evidently had other plans. I was to learn soon that the bureaucrats were extremely upset that the Prime Minister went to some place called Anand, met some fellow called Kurien and decided to create a National Dairy Development Board, when the Ministry of Agriculture already had a dairy division. They felt – fairly or unfairly – that by creating a dairy development board the Prime Minister was questioning the efficiency of the ministry's dairy division. This was viewed as an insult to the Ministry of Food and Agriculture.

Much later, when I got to know C. Subramanian better, he told me how officialdom had put up a bitter opposition to the Prime Minister's proposal for a national dairy board. The bureaucrats of Delhi believed that there was no need for such a body and if at all it had to be set up it must necessarily be located in Delhi. Subramaniam also informed me that seeing the truculent attitude of the bureaucrats, the Prime Minister told him that the project would work only if it was done in the manner that I had suggested, following the Anand pattern of cooperatives.

Within the brief two days that I was in Delhi I realised that I would not get even the modest sum of Rs 30,000 which I had requested. The bureaucrats put scores of conditions. Finally I met C. Subramaniam again and told him that I did not want to create any dairy board and it would be much better if I simply returned to Anand. I requested him to inform the Prime Minister that I had left.

The Minister tried to talk me out of it but when he saw that I was adamant he said: 'What shall I tell the Prime Minister?'

I was livid. 'You tell the Prime Minister that Kurien is not a beggar. He has not come here to ask for alms,' I said to Subramaniam. 'If you didn't want to give even Rs 30,000 you should have just said so. I would have managed without it.'

'Do you mean to say you can create the National Dairy Development Board without our money?' he asked.

I assured him that I could and left his office.

I returned to Anand and at the next board meeting of the Kaira Cooperative Union I confided to the members that I had a problem on my hands. I explained to them what had happened and said that the Prime Minister had asked me to create more Anands in the country but I would need their support and cooperation to be able to fulfil Shastriji's wishes. In other words, it was only with their help that I could set up the NDDB. Once more, I saw the stark difference between the way a government bureaucracy worked and the way our cooperative worked. It had taken two days in Delhi for the Agriculture Minister to try and get clearance for a small amount to start off a project which the Prime Minister himself was keen on initiating – and even he failed. Here, at the cooperative board meeting, in a matter of minutes we found a way in which the project could be financed.

There was only one concern that was voiced by one of the board members. He said: 'Why are you doing all this? Why create more Amuls? You will only create more competition for us.'

Whereupon I explained to the board what I felt – that one dairy for such a big country would not be enough in the long run. However big we thought Amul was, it would not be able to meet the country's needs. The country, as Shastriji had realised, needed several Amuls. Moreover, we should also keep in mind that one Amul was only one stick – it could be broken – a hundred Amuls together would be very difficult to break. Therefore it would be in Amul's interest to help create more of it.

Once the board was convinced with my reasoning, we worked out the modalities of creating the funds for the organisation and NDDB was set up. The Government of Gujarat gave us the responsibility to build three cattle-feed plants. We built them with Amul's help at a cost of Rs 30 lakh each. NDDB charged 5 per cent royalty to build these, so we collected a lakh and a half from each and were able to put Rs 4.5 lakh in NDDB's kitty. I was made NDDB's Honorary Chairman, another Amul employee was made the Honorary Treasurer and the third Honorary Secretary. Since we were already employees of the Kaira Cooperative, we received no salary from NDDB. The office space for NDDB – one room with a desk and a couple of chairs – was also within the Kaira Union premises. Almost everything that the NDDB got was provided at the cost of Kaira district's farmers. The technical and managerial competence of the Kaira Cooperative – the entire army of Amul – was placed at the disposal of the NDDB. In the early years, therefore, NDDB had an income but no expenditure. That was how NDDB was created – with absolutely no help from the government.

NDDB's brief was to replicate Amul – or the Anand pattern of milk cooperatives – throughout India. In essence, it meant that dairying had to be rescued from the government and given to the dairy farmers to whom it rightfully belonged. If by now my intermittent interactions with the government had revealed to me one truth, it was that the work of government should only be to govern. The government ought not to get into businesses. The government's rules and regulations were certainly not meant to run dairies; they were meant to administer the country. It was inevitable, therefore, that when the government tried to run dairies – as in the Delhi Milk Scheme or the Calcutta Milk Scheme or the Bombay Milk Scheme – they turned out to be utter disasters. This was the point that Shastriji understood and that is why he wanted NDDB to be created.

Once the Dairy Board was in place I thought that it was time to proceed to fulfil the mandate given to me. I began with the neighbouring state (by this time Bombay state had been divided into Gujarat and Maharashtra) and approached the Chief Minister of Maharashtra. What transpired was a fascinating process – a process that was followed like a pattern by most of the other states I approached. It provided me with an insight into the working of the bureaucrat's mind.

When I got to Bombay and asked the Chief Minister if I could create an Anand in Maharashtra, he seemed more than agreeable and he put me in touch with the Agriculture Minister. The Minister, in turn, arranged for a meeting between me and the Milk Commissioner. I took great pains in explaining to the Milk Commissioner what we were attempting to do at NDDB.

'So, according to you, the dairy should be owned by the farmers,' he observed. 'Then what should I own? You will dismantle my department. I know what you have done in Gujarat and there is no milk commissioner there – there is not even a milk department there.'

I said to him, 'I agree there's no milk commissioner or milk department in Gujarat, but there is milk in Gujarat.'

He was not moved and he told me that the farmers in Maharashtra were different, that my ideas were all wrong and while these strange ideas may have worked in Gujarat they would not work in other states.

I have never been one to give up easily and by now everyone who knew me knew that I thrived on challenges. Since it was well known that the logical and natural enemy of the Milk Commissioner is the person who holds the designation of the Director of Animal Husbandry, I decided to meet him and explain to him that the privilege of collecting milk from the farmers must carry with it the obligation to help the farmer increase production by providing him inputs. Anand at that time had seventy five veterinarians employed by farmers, with nine hundred first-aid workers. We had our own breeding stations; we did 300,000 artificial inseminations. I was quite sure he would be impressed by these figures.

His response was: 'Then what will happen to my department?'

I said, 'Surely, what happens to your department is irrelevant? What matters is how it benefits the farmers.'

Evidently, he did not think so and he threw me out saying that these ideas would not work there. Then I went to a third gentleman – the Registrar of Cooperatives. I explained the entire raison d'etre of Amul and NDDB to him as well.

'You are absolutely right Kurien. Cooperation is the only way,' he said. 'I have an excellent chap, a deputy registrar who is due for promotion. We'll make him the joint registrar and put him in charge of this project.'

By this time I was fast losing my patience. 'My friend, you have not understood anything,' I said. 'A cooperative is an organisation that is managed by farmers. It has to be run by the people employed by the farmers and not by government officers.' The Registrar of Cooperatives was not convinced either.

We approached state government after state government trying to impress upon them the need for milk cooperatives that would follow the Anand pattern. When the reaction from all state governments was practically the same – that my outlandish ideas perhaps worked in Gujarat but would not work in their state – I realised that it was not possible to do what the Prime Minister wished. To create Anands in the country would prove to be extremely difficult, only because the bureaucracy would not permit it.

In developing countries, one of the biggest obstacles is that the people do not have any power. It was quite clear that the state government officials were not going to permit the creation of dairies owned by farmers – particularly when I expected them to release funds from state resources. I then came to the conclusion that if we wanted to develop and create more Anands, we would need to have our own funds. Then I could go to the state governments and say to them: 'We will give you Rs 5 crore to create an Anand in your state but on one condition – that you ensure that village cooperatives are organised and milk is collected through cooperative structures; that a dairy is organised for the milk farmers and a genuine farmers' cooperative is established where the farmers will own the dairy.'

The seeds of Operation Flood began germinating in my mind.

~

Working from a small room in the Amul office, my colleagues H.M. Dalaya and Michael Halse helped me finalise a proposal that aimed to accelerate domestic milk production and that was economically sound and socially desirable.

Michael Halse was an FAO expert and a Harvard-trained Visiting Professor at the Indian Institute of Management (IIM) in Ahmedabad on a Ford Foundation grant. He used to visit Anand very often as he was researching on an agriculture project at the IIM and was keenly interested in our work with the dairy cooperatives of Kaira. We hit it off extremely well and when his term at the IIM was drawing to a close, I asked him to join us. But we could never have matched the money he was getting from the Ford Foundation so I came up with an idea. From the four million dollars we borrowed from the World Bank for Operation Flood, we gave a million dollars to the FAO, to get Mike for us.

Mike and I worked in close and harmonious coordination. Whenever I got an idea, I would call Mike. The two of us would sit and discuss it at length and Mike would take down extensive notes. Our discussions would often continue for a couple of days. Then Mike would disappear, often for four or five days. When he finally returned he would have a perfect, beautifully crafted document using my language exactly. This is how we produced the entire Operation Flood document between the two of us.

In many ways this marked the beginning of India's quest for food security in milk. It was no coincidence that the finalisation of our proposal occurred on 31 October 1968 – which was Sardar Patel's birth anniversary – and in Anand, so close to his birth place. Sardar Patel's birthday always marked the launching of an important project for the dairy farmers of Kaira, for in the late 1940s he had played a critical role in creating the milk cooperative in Anand and continued to inspire the farmers and their leaders with a vision that would ensure its success.

Some twenty years later the success of that cooperative, which had started with merely two cans of milk and a handful of milk producers, became the driving force behind a national programme to increase the country's milk production. Not many will remember this today – and indeed not too many in the government were willing to acknowledge it then – but there was an urgent need to initiate a national programme.

In 1950-51 per capita consumption of milk in India was 124 gm per day. By 1970 this figure had dropped to 107 gm per day, probably one of the lowest in the world and certainly well below the minimum recommended nutritional standards. Our dairy industry was struggling to survive. We produced less than 21 million tons of milk per annum despite the fact that we had the largest cattle population in the world. Things were beginning to look grim, particularly given the importance of milk and milk products in the Indian diet. When the NDDB team formulated a proposal for what we rather ambitiously named 'Operation Flood', it was hardly surprising that the government quickly overcame its earlier indifference to the programme and finally approved the proposal to enhance milk production through the cooperative method on a national scale.

To fully understand the need for a project like Operation Flood, it is important to know a bit about the patterns of cattle and milk movement in those days. All the research that our pioneering NDDB team did pointed to one extremely worrying trend. That was, that in the 1960s, the routine and senseless destruction of our high-yielding milch animals was intrinsic to the milk supply systems of our major cities. In Bombay and Calcutta, especially, the number of cattle kept in the city had increased dramatically during the preceding thirty years.

In Bombay, for instance, where there was not enough room for people to live, 100,000 buffaloes were kept in the most awful conditions. These buffaloes were brought into the city each year from the rural hinterland, as soon as they had calved. Some of them were brought from as far away as Punjab. The calf always accompanied its mother to encourage the rapid let down of milk. The first thing that they did in the city was to train the mother to let down her milk without suckling the calf. The process took about fifteen to twenty days after which, since there was neither the space nor the money to look after it in the city, the calf was mercilessly destroyed – frequently starved or drowned. Calf mortality was 100 per cent within twenty days of their arrival in Bombay. Thus 100,000 calves – the progeny of the finest buffaloes in the country – were killed in the city every single year.

As urban dwellings started expanding rapidly, cattle-keepers in these cities had to go further and further afield for supplies of green fodder. Therefore, they used only the highest yielding animals in order to get profit from their high cost milk production – and they found it unprofitable to freshen a milch animal when her lactation ended. What did they do with the mother buffalo? After seven to eight months the buffalo went dry, and if she did not conceive – which she rarely did in the city – she was sent off to the slaughter house. The younger ones were sent back to the villages miles away to calve again and produce more milk. The mortality of buffaloes in Bombay city thus was about 50 per cent each year, which meant that an astounding 50,000 of the best animals were destroyed in the city alone. Calcutta was no better except that there, since cow's milk was preferred for the famous Bengali sweets, the best cows and their calves would be taken to the city and then decimated.

The fallout from this destruction was unbelievable. In a wide arc – stretching from Gujarat, through Punjab and Uttar Pradesh – the best reserves of our milch animals were robbed of their high-yielding genetic material, as buffaloes, cows and their calves were herded into the cities over one or two short lactations and then thoughtlessly slaughtered.

Besides this, truckloads of cattle feed and fodder which had to be transported to the cities contributed to the increasing pressure on road and railway systems. Each khatal (stable) contributed its cattle's waste, dung and urine, to the cities' overloaded sewage systems and made the already precarious environment unhealthier.

Nor did the milk thus produced afford citizens much nutrition. Apart from the highly unscientific and unhygienic methods of milk extraction, the cities' milk producers found that their stable's supply of milk barely covered their costs, which included the extortionate rates charged by their moneylenders. Therefore, they would dilute the milk with water – usually impure water – and the consumer would get milk that was both adulterated and highly priced.

The government's liquid milk schemes (like the Bombay Milk Scheme), on the other hand, did not have the capacity to meet their city's entire needs. Moreover, as modern dairies, they could not dilute the milk they produced. For some time – when imported milk powder was cheap and the government spared the foreign exchange – they imported powder to subsidise their operations and marginally expand their meagre supplies of milk. Few milk schemes, however, could meet more than one-third of their city's needs and when they depressed prices by the use of imported powder, they discouraged local milk production. Efforts by most liquid milk schemes to increase their prices only led to private vendors raising their prices equally – so that, aided by dilution, they could continue to outbid the milk scheme for rural milk in the city's milkshed areas.

The ultimate loser was the common consumer and her children. In the cities, mothers saw milk getting increasingly thinner and more expensive each year, as the city got filthier and more unhealthy to live in. In the countryside, the ordinary milk producer saw her best milch animals going to the city for premature slaughter – while the milk that she produced from her remaining, lower-yielding buffaloes and cows still brought her only a small share of the rupee that the city consumers paid for that milk.

This was the anti-dairy development cycle that our proposal for project Operation Flood sought to reverse, by working out methods whereby milk would not need to travel to cities on the hoof. The project envisaged the organisation of Anand-pattern cooperatives in milksheds across the country from where liquid milk produced and procured by milk cooperatives would be transported to the cities. We expected this part of the project to cost to the tune of Rs 650 crore. We knew very well that the Government of India had many demands on its funds and if we were to ask for this amount for dairy development we were unlikely to get it. Therefore we evolved a unique methodology of funding our programme, to the extent of that Rs 650 crore.

Exactly at this stage there emerged a curious situation in the dairy industry of the developed world. For us, this turned out to be an extremely beneficial coincidence. In the late 1960s, there was a glut of milk products in the developed world. This presented us with a unique opportunity – an opportunity that was unlikely to recur, because the developed countries would be unlikely to repeat such a costly error, one which had produced such large surpluses of milk products. I realised that we at NDDB, as those responsible for dairy development in India must, therefore, quickly evolve a programme that would utilise these surplus commodities from Europe to generate funds required to finance the Rs 650 crore for Operation Flood.

Our proposal had the potential to not just stem the alarming decline in milk production but to reverse the trend to the advantage of both consumers and milk producers. The plan was to use donated milk products from overseas to facilitate protection of high-yielding cattle, resettlement of city-kept cattle and to obtain a commanding share of the market for the liquid milk schemes in four major Indian cities.

This was our project 'Operation Flood' which very soon came to be referred to as 'the billion-litre idea'.

At one level, the European Economic Community's (EEC) growing mountains of milk powder and lakes of butter oil could have posed a potential problem, which could have crippled our dairy industry. The European countries were already looking for a solution to their surpluses since these are not commodities that improve with age. I feared it was only a matter of time before some kindly European gentleman decided that the appropriate thing to do was to send those commodities to help feed 'India's starving millions' and, in doing so, crush for all time the growing aspirations of our dairy farmers. For if our government accepted this aid – as it surely would – and put it to consumption by poor people or children, it would thereby stimulate a demand without making any attempt to fulfil that demand with internal production thereafter. How could our farmers ever hope to compete with donated or concessional milk and butter oil sold to satisfy our urban consumers?

In every crisis, if you look carefully, you will spot an opportunity. My insistence on finding and seizing that opportunity has often been a source of annoyance for many of my colleagues because it means that unlike most people, I never try to sidestep a crisis. Rather, the more monstrous the crisis, the more I am tempted to rush at it, grasp it by the horns and manoeuvre it until it gives me what I want!

At this juncture, many enormous crises appeared on the horizon and I could only think of putting them all together and transforming them into an excellent opportunity. I convinced my colleagues that we must talk the European leaders into donating the commodities, but not simply to meet urban demand. We planned to convince them that they should gift the commodities to the NDDB, which would sell them at prices comparable to our farmers' prices, for reconstitution into milk and then for sale in our metros – again at prices comparable to our farmers' prices. We would thus utilise the donated commodities to build a market for quality milk in our cities.

However, it was one thing convincing my colleagues, quite another getting the government to appreciate and move the proposal forward. I had sent the Operation Flood proposal to the government with great enthusiasm and was quite crushed that there had been no response whatsoever from Delhi.

Then one day, the Inspector General of Police (IGP), Gujarat, Imdad Ali, telephoned me and said, 'Kurien, I am in trouble.' I jokingly said to him that if the IGP was in trouble, then all must be well in the state. He said, 'My big boss, Home Secretary L.P. Singh, is coming from Delhi on a visit. He's spending two days in the city. I have managed to work out his programme for a day and a half, but I find that there is still half a day left when he has nothing to do. Could I bring him over to Anand?'

I told him he could. I received L. P. Singh in Anand and showed him around the Amul dairy. He was amazed to see what the cooperatives had managed to do. I then took him home and there over a cup of coffee, the Home Secretary began to ask me why we did not have similar projects in the country, and why more farmers' institutions were not managed by farmers, and so on and so forth.

This was just the opening I was waiting for. 'It's because of your damned bureaucracy,' I said angrily. 'Because they don't ever recognise anything else done by anybody. They don't even know about Amul dairy. Your bureaucrats do not want experts. They don't want those who strengthen the people; they only want those who strengthen the bureaucracy.'

Then I told L.P. Singh that after the Prime Minister's visit and on his suggestion, I had already sent a proposal to the government explaining how Anand could be replicated in the rest of the country. But this proposal was gathering dust in the office of an officer at the Planning Commission. We had learnt that this officer was waiting for answers to two questions which were in his mind but which he had never asked! I had also sent a couple of reminders which were totally ignored. I told the Home Secretary that I was so fed up that I had finally quit trying. L. P. Singh asked if he could have a copy of the proposal and I handed him one which he took back with him to Delhi.

A few days later I got a call from him telling me that he had read the document and had spoken to the Cabinet Secretary, the Secretary, Agriculture and various other senior bureaucrats and he wanted me to come to Delhi to meet all of them. Still peeved that my proposal had been ignored for so long, I told him that I had no plans to visit Delhi anytime soon, but when I did, I would let him know.

After about a month, I had to go to Chandigarh for a meeting so I informed the Home Secretary that I was coming to Delhi at seven p.m. that evening but would leave for Chandigarh at six a.m. the next morning. 'I realise that we will have no time to meet,' I said, 'but since I had promised you, I am letting you know.'

When I reached Delhi at seven p.m. there was a car waiting for me at the airport. I was taken directly to L.P. Singh's house. He had managed to get all the senior bureaucrats to the meeting and they said, 'Well if Mohammed does not come to the mountain, then the mountain must come to Mohammed. Now tell us what is this Operation Flood.'

I spent three hours with them. L.P. Singh looked at the gathering and said: 'He has asked nothing from us, other than to forward his proposal to Rome. How come we have not done this yet? He has already spoken to the executives at the World Bank, he has spoken to the Director-General of FAO and so on, and we have just been sitting on it? What kind of bureaucracy are we heading?'

The bureaucrats immediately told me that they would hold a Committee of Secretaries meeting attended by the Cabinet Secretary, Shivaraman, and I should also be present because they would approve the proposal at this meeting. This is how Operation Flood was finally approved by the Committee of Secretaries and sanctioned by the Government of India. No politician was involved in this sanction. When I went to Rome to present the proposal at the meeting of the twenty-two nations, B. R. Patel, the Agriculture Secretary, accompanied me.

The proceeds from this project ultimately financed the creation of a cooperative structure in India which, as I write in 2005, involves around eleven million dairy farmers in twenty-four states of the nation. These farmers today own their own dairy plants, are members of cooperatives, cooperative unions and federations and market some of India's best-known dairy products.

OPERATION FLOOD

THE YEAR 1965 WAS FAIRLY MOMENTOUS FOR ME. THAT YEAR, MICHIGAN State University conferred on me an honorary degree of 'Doctor of Science' amid great fanfare, thereby promoting me from an ordinary 'Mr Kurien' to 'Dr Kurien'.

In September 1965, NDDB was registered as a society under the Societies Registration Act, 1860. My colleagues and I, energised with our grand scheme for Operation Flood, impatient to implement it, were thinking of ways and means of receiving the gifted commodities and utilising them to generate funds for various activities within the project. Our proposal envisaged in detail how the European surpluses were to be used to speed up India's dairy development.

In a nutshell, our approach involved three basic steps. The first was that the donated milk products would be reconstituted, in order to provide the Bombay, Delhi, Calcutta and Madras liquid milk schemes with enough milk to obtain a commanding share of their markets. Next, the funds realised from this reconstitution and sale of donated products were to be used to resettle city-kept cattle and help them to breed and to increase organised milk production, its procurement and processing. Finally, this entire operation would be directed towards stabilising the position of major liquid milk schemes in their markets.

We sent in our proposal to the World Food Programme (WFP)

for approval. As we eagerly awaited a response from them, we had a reason to worry. Kesteven, an Australian, Chief of the Animal Production and Health Division of the FAO, warned us that unless we got our proposal through very fast, we stood to lose it. He confirmed to us what we already knew – that the European Union was in a hurry to utilise the food aid. But he also told us something we did not know.

Kesteven confided to us that he had heard whispers that Nazir Ahmed, a Pakistani who was the head of the WFP for this region, wanted the aid to go to Pakistan. Nazir Ahmed had copied our entire proposal and sent it to the Ambassador of Pakistan in Rome, advising him to simply substitute the names of cities – change Bombay to Lahore, Calcutta to Karachi and so on – and submit it as Pakistan's proposal to the WFP.

I was aware that Nazir Ahmed was deeply prejudiced against the Indian government. I remember that he had once asked me how a Christian like me could be designated Chairman of NDDB. I had replied: 'Mr Ahmed, that is because India is not Pakistan. When your country attacked India, the Collector of Kutch district was a Christian, the IGP in Gujarat was a devout Muslim, the Home Secretary of Gujarat was a Christian and the Governor of Gujarat was a Muslim. That is India for you.'

Kesteven now urged us that if we did not push harder and faster, the aid – or at least half of it – might go to Pakistan. However, Nazir Ahmed's ploy did not work and I was asked to meet a committee of the WFP to formally present the proposal for Operation Flood. With no help from the government, NDDB planned the programme and negotiated the details of EEC assistance.

I can still recall my visit to Rome in October 1968 to present NDDB's project proposal to a twenty-four nation executive committee of the WFP. The Secretary, Agriculture, of the Government of India at that time was B.R. Patel, later the

Vice Chairman of Air India. As soon as he met me in Rome he said to me, 'Look here, I don't know anything about any of this so you have to handle it all. My job would be entirely limited to saying: "India's point of view will be presented by Dr Kurien." And with that I will end and you will take over.'

One thing I abhor is delivering speeches from prepared notes. Very often colleagues have helpfully tried to present me with notes for my speeches but once I stand up to face the audience, the need to share my feelings and thoughts leads me to quickly put those notes aside and speak extempore. I had not prepared a written presentation for this gathering in Rome either. Yet, I was so completely immersed in the concept of Operation Flood and so convinced that I delivered an intensely impassioned speech, completely off the cuff but straight from my heart. I must have given a pretty good performance and my arguments must have been convincing, for I received an astounding response.

The impact of my presentation was tangible, for almost all those present came up to congratulate me. Two responses in particular delighted me. The head of the Australian delegation shook my hand warmly and said: 'I came here determined to oppose your project but after you spoke, how could I?' And I felt another sense of victory when a Pakistani delegate came to me and said: 'Dr Kurien, my brief was very simple: "Oppose everything that India proposes". But after you spoke, I simply could not do that. So please understand that my silence is my support.'

Essentially, the focus of my speech was on how food aid, when used judiciously, could become an instrument for social and economic well-being and not something that would mar and destroy that country's self-sufficiency. I convinced the delegates that being an employee of the farmers of India I would use this aid for the good of my country, not for the good of the EEC. I warned them that I would not allow them to create by their donations a permanent market here – I would not allow them to 'dump' their dairy products in my country. And because I recognised that food aid is a two-edged sword I would not wield it unskillfully, thereby cutting our own throat.

I informed them that, in fact, in India one-third of the milk being supplied in the cities of Bombay, Delhi, Calcutta and Madras was from imported powder but I was the only one who spoke of it. Even baby-food manufacturers used imported milk powder. I told them bluntly that I recognised that food aid was an 'investment' by donating countries .

If a donating country gives us $100 million worth of milk, they are, in fact, investing $100 million in the country at a time when there is a world surplus in milk and global prices are down. I was aware that they were doing so with an intention, hoping that it would have a residual impact – that is, if they distribute $100 million worth of milk, at least $10 million residual demand would remain in my country, which would, perforce, have to be met by imports. Therefore, they expected a 10 per cent return on their investment, which was not a bad return.

Moreover, since they were giving the aid at a time when the prices were down and later, when India was required to buy the commodity, the prices would have shot up, that 10 per cent could have actually become 20 per cent. I told them that donating countries were well aware of these tricks when they gave food aid. They would, in fact, like to see India, an important milk-producing country, buckle under its feet.

I assured them that it was my job to see to it that this did not happen. I promised them that I would not allow India to become dependent on their gifts; I would not allow these products to be distributed in my country at a price lower than what Indian producers could demand; I would not allow it to become a disincentive to produce milk here.

My government had nominated me – as Chairman of NDDB – to receive these gifts because I was aware of what needed to be done to protect the interest of my country. Since I served the farmers, the milk producers of India, I would never allow the largesse to damage the interests of the Indian farmers. I convinced the gathering that the food aid would be handled differently and would not be distributed at low prices in the ministers' constituencies to make them popular – not even in the constituency of the Prime Minister. Above all, I hoped that they would understand that this project was workable because it would not have a bureaucrat at the helm.

Having said all this, I also stressed that it was crucial that India receive the aid because it would give us an opportunity to replicate the Anand pattern at the national level. The money so generated would be used solely to stimulate production and create a structure that would ensure a continuity of this process. Without their cooperation, I pointed out, the project would remain a mere dream.

After a prolonged session, Nazir Ahmed felt he must have the last word. He asked, 'How can we be sure that your proposal is foolproof?' There was silence among the audience. I then replied, 'I thought I had made it foolproof but I see that I have failed.' After a long silence everyone burst out laughing.

Clearly, my arguments were convincing. The proposal for the donation of food aid from Europe to India was approved by the WFP and the project agreement was signed between the Government of India and the United Nations/WFP in March 1970. It had taken over a year from the time my colleagues and I first conceptualised the project, but now we had won our first battle to initiate Operation Flood.

At first, NDDB was to be the operating authority for the project, handling the donated gift commodities, selling them to the dairy plants and using the funds thus generated for the implementation of Operation Flood. However, it was felt that since NDDB was merely a registered society, it would not be able to cope with such a massive task. Therefore, in February 1970 the Government of India set up a new public-sector company, the Indian Dairy Corporation (IDC), under the Indian Companies Act of 1956, to receive the gifts, generate funds by their sale and disburse the funds to implement Operation Flood. NDDB was to continue as a technical consultancy body, responsible for preparing feasibility studies, undertaking dairy projects on a turnkey basis and so on. The government appointed me Chairman of IDC too.

In July 1970, NDDB officially launched 'the billion-litre idea' with its goal to take India's dairy industry from a drop to a flood. It was to become the biggest dairy development programme in the world.

~

The purpose of Operation Flood, of course, was to replicate Anand in a number of India's major milksheds. The basic principles underlying this project are so simple that many people tend to overlook them. In hindsight it amazes me how all the dairies built in India before Operation Flood ignored them.

The first of the principles was that if we wanted to produce milk, we must have a market for it and that market must be such that the person producing the commodity must profit. After our experience at Amul, we knew that we could depend on our farmers. The farmers have a long and ancient tradition behind them and a great wisdom born out of experience, and once we assure a market and a good price, they will produce.

Unfortunately, our directors at animal husbandry and our directors of the milk department did not understand this fundamental principle whenever they talked of increased production; they spoke endlessly of better seeds, fertilisers, irrigation or of vaccines, Holstein semen, Jersey semen, warm semen or frozen semen and so on. None of these people ever spoke about marketing, which was so central to increased production. We at NDDB knew that there would have been no Anand unless there was a Bombay. There could be no production unless there was a market. Marketing, therefore, was the dominant orientation of the project as a whole.

Another lesson we had already learnt at Amul was that we could not link procurement and marketing rigidly. It was true that in the case of milk we could not sell more than we collected and neither could we sell less, and therefore, the market needed to expand just that much more. However, if we had linked procurement and marketing rigidly, we would have created a situation where markets had to expand at exactly the same rate as production or procurement – which is impossible. What, for instance, would happen in summer, when the production and procurement of milk goes down? The market would not shrink to suit us. So we knew that the moment we linked the two rigidly, things would not work.

The second basic principle was that an Anand could not be created next to Bombay. For instance, we knew that if we had tried to collect milk from within a radius of thirty miles of Bombay, as the Delhi Milk Scheme tried to do in Delhi, it would be doomed because then all you need is a doodhwala bhaiyya – a local milkman – on a bicycle. Expensive structures like the Delhi Milk Scheme, with thousands of employees and the IAS men on top, were unnecessary to merely carry milk some thirty miles into town to the consumer. With their overheads, they could never have hoped to compete with the doodhwala bhaiyya and won. The local milkman adds water to the milk every summer when the milk supply drops; when in winter he has too much milk, he refuses to collect more than he wants from the farmers. Government departments, naturally, could not do all this. Thus, while government dairies continued to collect, pasteurise and distribute a certain quantity of milk daily, they did so not as viable units but as loss-makers depending on heavy subsidy.

We at Anand knew that if we wanted to get hold of procurement, not only must we fight with those who collect milk – the traders, the milk merchants – and persuade the farmers to give the milk to us, but we must also attack those same milk merchants in the areas where they sold the milk. We had already done this in Baroda. The cooperatives in Baroda could not capture the available market in summer because they did not have enough milk. In winter when they had all the milk, there was no market. When the Chairman of the Baroda Cooperative came to me with this problem, I proposed to him a cooperative solution. I said that the Kaira Cooperative would give them 25,000 litres of milk from Anand in summer, so they could capture an additional 25,000 litre market. In return, they could send us their excess supply in winter. When we did this, the Baroda Cooperative found that in summer their collection automatically went up by 25,000 litres because their competitor lost that much in the market.

Through Operation Flood we would get commodities from Europe and put that milk into distribution, capture the market in Delhi and the other cities, whereby the merchants would lose the market and the collection of milk would fall into our laps. These, then, were the simple principles underlying this project, most of them based on good sound marketing. The bedrock of all this, naturally, would be the cooperative system, which would not provide any room for those sucking away margins exploiting both the producer and the consumer. Operation Flood would, therefore, help in establishing a direct link between those who produced the milk and those who consumed it, and the cooperative structure would ensure the transfer of the maximum share of the consumer's rupee to the producer.

When my colleagues and I designed and proposed Operation Flood, in our minds the project had other, larger implications too, besides achieving self-sufficiency in milk. We also wanted to ensure the development of women and men belonging to one of the most neglected sections of our society – our farmers. We were convinced that this could be done only if we involved them in the process of development. When the British ruled us they said why give these 'natives' any freedom, they will only make a mess of it. Similar was the 'brown sahib' mindset. It was strange and ironic that we forgot how to trust our own farmers despite the fact that our own ancestors must have been farmers.

Knowing the attitude of our brown sahibs, I was expecting opposition and was hardly surprised when it came as Operation Flood began, because dairying was already dominated by governmental structures. One hundred government dairies had already been built in different parts of the country. Whenever there was a problem a department was created to look into it, a commissioner was appointed and a joint commissioner immediately followed. A whole cumbersome hierarchy was created, in keeping with the bureaucratic approach in India.

On the other hand, Operation Flood was to be completely decentralised. Village matters would be handled by village societies. Milk would be collected by people of the villages; at the district level the dairies would be handled by the district unions; at the state level the marketing would be taken care of by the marketing federations. In our dairy cooperative structure, the production of milk is kept in the hands of the small farmers, the marginal farmers and landless labourers. The entire structure, therefore, fits in with the spirit of India, of a decentralised, rural economy. The cooperative structure never encourages huge, bureaucratic systems, for it knows that mammoth bureaucracies cannot be sensitive to the needs of people. Unfortunately, our bureaucracy has been against such decentralised systems.

Naturally, therefore, we had difficulty in getting such structures accepted by certain bureaucrats. In many states there already existed dairy development departments employing 25,000 people. These departments imagined that they existed in order to give jobs to these people. I believed they existed only partly to give jobs and partly to provide marketing facilities to millions of farmers. This was a fundamental difference in our approach and problems and conflicts arose constantly.

'Where cooperation fails, there fails the only hope of rural India'. Interestingly, this was a statement made by the Royal Commission of Agriculture before independence, but very few in our country recognised the truth behind it. Unfortunately, as things stand today, many Indian cooperatives do not function efficiently because they have been made official, bureaucratised, politicised and therefore, they have been effectively neutralised. If, on the other hand, these cooperatives are restored to the farmers, and the farmers are left free to manage them, they will become truly democratic structures.

~

Operation Flood made our country self-sufficient in milk and this was achieved entirely through the cooperative structure. Today, eleven million farmers in twenty-two states across the country own two hundred dairy plants handling eighteen million litres of milk a day. This is a remarkable achievement. While we in India tend to take our achievements for granted, this feat elicited high praise and admiration throughout the world. It did, however, take decades of sustained, hard work.

In 1955 our butter imports were 500 tons per year. Today, our cooperatives alone produce 12,000 tons of butter. Had it not been for Amul and then Operation Flood, it is estimated that butter imports would have gone up from 500 tons to 12,000 tons. Similarly, we imported 3,000 tons of baby food in 1955. Today our cooperatives alone produce 38,000 tons of baby food. By 1975 all imports of milk and milk products stopped. The only import permitted was that of food aid under Operation Flood – 37,000 tons of milk powder and about 11,000 tons of fat. That was about one per cent of the milk production of India. To proclaim – as our critics did – that this one per cent would make us dependent on food aid did not make sense. We could have done without that one per cent. But that one per cent was important to us at NDDB/IDC because it generated the money we needed and set the stage for marketing, which was necessary to stimulate dairy development in our country. It helped us create the national milk grid through which milk from the rural milksheds was transported to seven hundred towns and cities. It enabled us to create this large cooperative structure through which millions of dairy farmers and their families benefited.

Having ensured that the food aid from Europe would come to India, some other systems needed to be put in place internally if the entire operation was to work. One of the issues I discussed with the agriculture ministry was that if I, as Chairman of NDDB and IDC, were to guarantee the success of Operation Flood, if I were to be responsible for increasing the milk production of this country and developing its dairy industry, then NDDB/IDC would have to be the official canalising agency for the import of milk powder. I had seen that in New Zealand there was only one seller of milk powder – whoever wanted milk powder, had to go to the New Zealand Dairy Board. We figured that similarly, if there were one buyer in India, then the scales would be more evenly balanced. The Government of India agreed to this request but this too led to a huge controversy.

Much of the controversy arose because people did not fully understand the role of a canalising agency. The common conception was that as the canalising agency, NDDB would decide how much of the imported milk powder should go to whom. This was not true. A canalising agency merely canalised the import of the commodity. Those days the Ministry of Industrial

Development decided how much milk powder should be given to baby food manufacturers, how much should be given to malted food manufacturers, to condensed milk manufacturers and so on. This ministry provided NDDB a consolidated list of who was to get how much. The Ministry of Agriculture informed us how much should go to the various city liquid milk schemes and gave us a list for that. The job of NDDB as the canalising agency was only to import milk powder for India on the best possible terms and distribute it as per the government's allocation.

We performed both tasks diligently. As the canalising agency I think we struck some excellent deals with milk-producing countries. When I went to New Zealand, I met my counterpart there, the Chairman of the New Zealand Dairy Board, and I told him that I wanted to buy some milk powder and that they should sell it to me at the same price they sold to England. After some bargaining, he agreed. Then I suggested that the freight be half of what it was to England since the distance was half. It took far harder bargaining to push that through. Having done that I said to him that if he wanted me to buy milk powder, he should also gift me some for we were, after all, a developing nation. Since this was a government decision he called the Minister for Agriculture over lunch and I convinced him that if the Government of New Zealand donated 5,000 tons of milk powder, we would buy another 5,000 tons. Once they agreed to this I convinced them to continue this agreement for three years. They agreed. Then I went to Australia and I talked to their minister. After a couple of days of intense bargaining, they, too, agreed to the same terms for one year. In effect, thus, I managed to get the imports for our country at half the price. That is what a good canalising agency does.

We were equally efficient on the distribution side. I had already told the Government of India that while NDDB would give imported milk powder to anyone that the government allocated, if we suspected anybody on that allocation list we would inform the government and take steps to clean up the corruption. The government had no doubt about our intentions – to choke out imports of milk powder within four years – and they agreed. This was what was done and it was a great achievement.

At one time, NDDB received a list from the government which had listed 276 baby food factories for allocation in Karnataka. I knew that Karnataka had, at best, one baby food factory and so I sent one of my officers to investigate. He came back with the information that 275 of these so-called factories were merely name boards pasted on walls without even a single office to that name. This had become lucrative business. Dummy factories were registered and unscrupulous businessmen got import licenses worth 10 million rupees. They sold these for 30 million rupees and pocketed 20 million rupees. They never even touched the milk powder.

I asked the Ministry of Agriculture to inquire into this list. I warned all these so-called baby food and condensed milk manufacturers: 'Gentlemen, the game is up. Now you're talking to a dairyman, not to the wise people of Delhi, so you must be careful. You were given a license in order to make baby food and other dairy products from Indian milk, not in order to import milk. From now on, your import license will be cut by 25 per cent each year so you have exactly four years to get milk produced in the country.'

Whenever imported milk powder was allocated, we informed the manufacturers that this would be given at a price which would be at least the price it would cost them if they bought milk in India. They would get no subsidy and we warned them that imports would be cut by 25 per cent each year. So whether it was the Bombay Milk Scheme, the Delhi or Calcutta Milk Scheme or Nestle or Glaxo, they had to start procuring all the milk they needed within the country in four years. I had the full support of government at the highest level to do this and, therefore, I could confidently tell the manufacturers that no matter how many strings they pulled they would not get imported milk powder beyond four years. If anybody needed proof of NDDB's integrity in this, it was evident in the fact that one of those who had their import license cut by 25 per cent each year was Amul because, like other baby food manufacturers, Amul, too, held an import license.

These strict rules that we adhered to as the canalising agency created a lot of waves. Many felt that the government had invested NDDB with far too much power. Some of the manufacturers supported me; some of them opposed me. But I had laid down the terms and conditions with the approval of the Government of India, which recognised that during Operation Flood I would need to do all this. I have to reiterate that this was possible because of Shastriji's visit to Anand and subsequently his total support to the project and his complete faith in us. After Shastriji's tragic death, Indira Gandhi became the Prime Minister and she, too, was also extremely supportive. By then I was already on to a good wicket!

Did all this really invest the NDDB with too much power? Perhaps it did. But it was a power that we exercised prudently over those who tried to expand our country's imports unnecessarily; it was a power we exercised over those who, by their corrupt practices, held the country back from true development. We succeeded in the dairy sector because we were given the power to stop them. But we did not cut their throats ourselves. We were clever enough to get Prime Minister Shastri and his government – and subsequent prime ministers and their governments – to cut the throats of the corrupt!

I have always maintained that since all the selfish and ambitious people from our villages constantly make a beeline for Bombay, Delhi and other glamorous cities, the people who are left behind in our villages are only the good people. It is necessary that we mobilise these good people and involve them in the processes of development. Operation Flood was one project that placed its faith squarely in these people of rural India. What we needed to do was to organise the farmers, place them in command and be their employees to provide the technical, administrative and other strengths. This unbeatable combination at NDDB is what became the organisation's irresistible force.

When Operation Flood was sanctioned I knew that it was a massive and extremely complex operation and we would need all the help we could possibly get from all quarters. It was in this connection that, one day, I called on J.R.D. Tata, Chairman of one of India's largest industrial houses, one known for its commitment to quality and for its patriotism. I met him and explained to him the entire concept behind Operation Flood. I told him that such an enormous task would be extremely difficult to pull off alone and I requested him to spare six managers from the house of Tatas for one year, to help us improve the nation's dairy industry. I could pay them only public-sector salaries, but within that, I assured him, I would pay them the best that I could. At the end of that year, his managers would return to his company, far richer for their thorough understanding of cooperatives and of agriculture. I was confident that it would be an extremely valuable experience for his managers.

J.R.D Tata listened to me very patiently and then told me that since this was not a decision he alone could take I would have to present it to the board. I agreed to do so and met the board and once again explained the intricacies of the entire project to the members. They, too, listened very politely, smiled and nodded. But that is as far as they were prepared to go. To this day, I do not know whose decision it was, but we were loaned not even a single manager from the Tata Group. After all, would it have so adversely impacted the Tatas if they had deputed six managers to the NDDB and that, too, for a brief period of one year?

The incident left me with a bitter taste and justified my belief that, in the ultimate analysis, the corporate world and the

Left:
An all-rounder: Verghese Kurien during his college days.

Below:
Dr Kurien outside the garage which was his first home in Anand.

Left:
Dr Kurien with his parents, brothers and sister.

Below:
Soulmate: Dr Kurien with Molly on his wedding day.

Right:
A family man: Dr Kurien with wife Molly and daughter Nirmala.

Below:
Dr Kurien and Molly with Maniben Patel, who was like a surrogate mother to them.

Left: (Left to right)
Support structure:
Dr Kurien with Morarji Desai, Tribhuvandas Patel, Nirmala, Maniben (Tribhuvandas's wife), Molly and Maniben Patel (Sardar Patel's daughter).

Below: (Left to right)
Friends and family: Tribhuvandas Patel, Zarine Variava, R.H. Variava, Maniben Patel, Yasmin Variava, Nirmala, Molly and Dr Kurien.

Right:
Apple of their eye: Dr Kurien and Molly with grandson Siddharth.

Below:
Agent of change: Dr Kurien with Molly and Nirmala.

Left:
Humble beginnings: The place from where Amul originated.

Below:
Dr Kurien with Prime Minister Lal Bahadur Shastri at a milk collection centre.

Right:
Solid foundation: President Rajendra Prasad lays the foundation stone of the Amul dairy.

Below:
The first Amul dairy.

Left:
Dr Kurien with Morarji Desai, a firm support behind the cooperatives.

Below:
Dr Kurien with Michael Halse, who helped him formulate the plan for Operation Flood.

Right:
(Front row) Colleagues at Anand: R.H. Variava, H.M. Dalaya, V. H. Shah *(Back row)* Dr Kurien and Pheroze Medora.

Below: Dr Kurien receiving the Ramon Magsaysay Award in 1963 from the President of the Philippines.

Top left:
Dr Kurien receiving the Padma Bhushan in 1966 from President S. Radhakrishnan.

Top right:
Dr Kurien receiving the Honorary Doctorate of Law from Glasgow University in 1974.

Left:
Dr Kurien receiving the Krishi Ratna Award from Balram Jhakar in 1986.

Above:
Dr Kurien at the XIX International Dairy Congress (1974) with Jagjivan Ram and Faqruddin Ali Ahmed.

Right:
Dr Kurien receiving the Wateler Peace Prize (of the Carnegie Foundation) in 1986.

Left:
Dr Kurien receiving the World Food Prize in 1989.

Below:
Dr Kurien receiving the Padma Vibhushan from President K.R. Narayanan in 1999.

Right:
Dr Kurien receiving the Wisconsin – Dairy Man of the Year Award in 1993.

Below:
Dr Kurien as the chief guest at President A.P.J. Abdul Kalam's doctorate ceremony in 2000.

Right:
Dr Kurien with HRH Queen Beatrix of the Netherlands.

Below:
Dr Kurien with Soviet Premier Alexei Kosygin and George Fernandes.

Facing page: Top left:
Dr Kurien with the Aga Khan.

Top right:
Dr Kurien with Prime Minister Indira Gandhi.

Below left:
Dr Kurien with Prime Minister Jawahar Lal Nehru.

From an organisation to an institution: Amul dairy I and II.

cooperative world are distinctly different. I decided that we would find and employ our own managers and that is what we did. It took us a while but we got excellent people to join us – people of commitment and integrity – and perhaps, in the long run, that was the better way of doing things. Together, we implemented Operation Flood and we implemented it well.

When a developing country starts to become self-sufficient, naturally, there are many who are affected by it because over time, the outflow of foreign exchange stops. Our dairy cooperatives were entirely Indian, run with Indian technology. Whenever we needed some expertise we asked an organisation like the FAO to help us. We did not give them participation in share capital and repatriation of profits. There came a time when just one district cooperative – the Kaira District Cooperative – became bigger than any multinational operation in India. Opposition was, therefore, expected from multinational lobbies.

What was more painful to swallow was that within my own country, too, there were those who could not stomach the common man becoming stronger. The existing power structures did not like it; the merchants certainly did not like it. Naturally then, there was opposition from within the country too, but we understood that when people attacked us at NDDB they were actually trying to attack the cooperative structure and undermine it. However, such was the momentum of the cooperative movement in the dairy sector that the flood could not be stemmed. If the vested interests had wanted to stop it the only thing they could have done was to have destroyed the first Chairman of Kaira Cooperative way back in 1945. When Tribhuvandas Patel started the first cooperative the farmers realised their potential and once they came into their own, they soon reached a point when nobody could stop their growth.

Having seen the success of the Anand pattern, we at NDDB were determined to overcome all obstacles and stoically face whatever criticism came our way in order to carry the vision of Anand to the rest of the country. We strategised, we got positive thinking people on to our side. We helped them understand that only cooperatives would bring power to their people and, therefore, prosperity to the state. Our efforts began to bear fruit. Many of the states, which earlier threatened us that Operation Flood would come to their states over their dead bodies, had to start digging their own graves. Karnataka, for instance, which had been extremely hostile to Operation Flood, finally had to face a situation where sandals were thrown in the Assembly and the government was asked why Operation Flood was not being brought to their state.

Operation Flood eroded the authority of many agencies and departments and this is why it was so hated by those in power. Whereas those in authority resisted it, all the farmers from Kashmir to Kanyakumari were enthusiastic about it, for they found themselves in a similar predicament – they were helpless, exploited and they needed our help. Our spearhead teams went from village to village explaining to them what the cooperative structure and Operation Flood meant. We brought groups of farmers to Anand to show them how the experiment had worked. We did this with bureaucrats from various states, too, and the reaction of most people – whether farmers or bureaucrats – was a sense of incredulity at the fact that something like this could happen in India.

When almost five years had passed since we launched Operation Flood, we felt it was important for us to detail the progress of the experiment. Already a few documentaries had been made and released. One day, film-maker Shyam Benegal (then with Radeus Advertising), who had made some of these documentaries, pointed out to me that there were a great many fascinating and moving human stories involved in the setting up of the cooperatives and the implementation of Operation Flood. He felt that something more needed to be said and while documentaries were effective and they gave the figures and the information, the genre could not do justice to the human stories that he saw emerging. He wanted to record the changes in human beings in our villages and felt that documentaries would be a bit too dry for this.

I suggested to him that he should think of making a feature film and he agreed. There was the small issue of money, however. How would we raise the funds needed to finance a full-length feature film? I consulted my colleagues and we came up with the idea to ask all the members of Gujarat's milk cooperatives to contribute money. Every single milk producer contributed a rupee each and we managed to raise the necessary funds. Thus, in effect, the film, *Manthan,* was produced by Gujarat's dairy farmers. It was a rare, low-budget film, which cost only a million rupees to produce. Moreover, in the highly commercialised Hindi film industry notorious for its black money transactions, the funds for *Manthan* were paid completely in white from a bank account held jointly by Gujarat Cooperatives Milk Marketing Federation (GCMMF) and Radeus, who were the executive producers. The accounts were fully audited.

I have always believed that once you identify the best person for a particular project and tell him or her exactly what you expect, you must put your complete trust in that person, allowing him or her to work independently without interference. If you do, the project is bound to succeed. This was a lesson I learnt very early in my career from Tribhuvandas, who put his entire faith in me and allowed me to work as I saw fit. I applied the same logic of delegation in many spheres of my own work. When we had to build the NDDB and IRMA campuses, I thought of one of our best architects, Kanvide, to handle the projects; when we decided to set up our own school, Anandalaya, I asked Vibha Parthasarthy of the Gujarat Education Society and at that time the Principal of

Delhi's Sardar Patel Vidyalaya to help. It was the same with the choice of Shyam Benegal. He had seen the cooperatives grow over many years. He had been involved in one way or another with the Anand experiment – first as an ad film-maker for Amul and then with NDDB's documentaries on Operation Flood. Who else could have been a more appropriate choice to make a feature film on the human dimensions of Operation Flood?

Shyam Benegal did not let us down. He gathered a team of extremely talented and dedicated professionals and they began their work. The renowned playwright, Vijay Tendulkar, spent many weeks in Gujarat's villages researching and putting together the story, culled from the experiences of many farmers of many villages. It took the team some nine months of actual shooting in various parts of Gujarat to complete the film. As Director, Benegal assembled an extraordinary cast of young actors including Smita Patil, Girish Karnad, Naseeruddin Shah, Anant Nag, Kulbhushan Kharbanda, Amrish Puri and others, all of whom went on to very distinguished careers in Indian cinema. A superbly directed and acted movie, *Manthan*, which was released in 1976, was a commercial success, winning several awards.

More than anything, however, *Manthan* became an invaluable part of the kit carried to the villages by our spearhead teams who used it – first in super 8 mm cassettes and later as videos – to educate our farmers about the difficulties and rewards of cooperation. All across India, cooperative societies bought the film to show to their members. The message of the economic and social benefits of the cooperative structure could not have been more vividly and effectively depicted.

While conceptualising the ambitious Operation Flood had been relatively easy, its implementation proved to be a gargantuan task. We decided that it would have to be executed in three practical and workable phases. Ultimately the three phases spanned almost three decades.

Phase-I, which we had hoped to complete in five years, finally took twice as long, spanning the 1970s. This, as we had meticulously planned, was financed by the sale of skimmed milk powder and butter oil gifted by the EEC through the World Food Programme. Through Operation Flood-I we linked eighteen of India's premier milksheds with the consumers of our four major cities – Delhi, Bombay, Calcutta and Madras.

Work proceeded at a frenetic pace and we needed all the help we could get if we wanted to avoid further delays. One of the constraints placed on us was by the government's own bureaucratic working. In 1977, I finally approached Prime Minister Morarji Desai with my tale of woes. I told him that getting any NDDB file passed had taken on a predictable, frustrating pattern, which went something like this: I would go with a file and meet the secretary, who after some discussion would tell me, 'Yes, government will look into it.' When I asked him who this 'government' was, who 'would look into it', he would tell me it was the minister. So I would go to the minister with the file – first the Minister of Agriculture, then Finance, Planning and Environment – where once again the issue would be discussed threadbare and I would be told: 'Yes, the government will look into it.'

Once more I would be forced to ask honourable ministers who this 'government' was and this time I would be told it was the prime minister. I would take the file and meet the prime minister to discuss it again, only to be informed for the umpteenth time, 'Yes, the government will look into it.' By this time, totally nonplussed and thoroughly exhausted, I would ask the prime minister who was this 'government' that would finally take a decision on my file? And I would be told it was the Cabinet. By the time this tedious, excruciating merry-go-round was completed, a whole year would have gone by and I was left with the nasty feeling that if each person I met could only tell me that 'the government will look into it', it probably meant that no one would look into it. Each one of them was only a file pusher, too scared to take any decision, God forbid if that decision turned out to be the wrong one. How was it possible to get any work done this way?

Morarjibhai smiled wryly, hearing my sorry tale. But for the first time, he asked for a committee of secretaries to be formed, which would specially handle all NDDB's papers. With this new system in place, once the Ministry of Agriculture had seen the file, it would automatically go to the committee and if they had any questions, NDDB officers would meet them and sort out matters right away. Mercifully, the merry-go-round came to a halt.

The second phase of Operation Flood, which lasted from 1981 to 1985, was implemented with the seed capital raised from the sale of the EEC gifts as well as a World Bank loan of Rs 200 crore. With this phase, the number of milksheds increased from 18 to 136. Moreover, 290 urban markets increased the outlets for milk produced. By the end of this phase, a self-sustaining system of 43,000 village cooperatives covering 4.25 million milk producers was well established. Milk powder production increased from 22,000 tons in the pre-Operation Flood year to 140,000 tons by 1989. All this increase came from dairies set up under Operation Flood. Direct marketing of milk by producers' cooperatives increased by several million litres a day.

The third phase of Operation Flood, which lasted from 1985 to 1996, added 30,000 new dairy cooperatives to the 42,000 existing societies. Member education was intensified, and significantly, the number of women members and women's dairy cooperative societies increased considerably. This phase focused on assisting unions to expand and strengthen their procurement and marketing infrastructure to manage the increasing volumes of milk (by 1989 the number of milksheds had grown to 173). Veterinary health-care services, feed and artificial insemination services for cooperative members were extended. During this decade we placed increased emphasis on research and development in animal health and animal nutrition.

In 1998 the World Bank published a report on the impact of dairy development in India and looked at its own contribution to this. The audit revealed that of the Rs 200 crore the World Bank invested in Operation Flood, the net return into India's rural economy was a massive Rs 24,000 crore each year over a period of ten years. Certainly no other development programme, either before or since, has matched this remarkable input-output ratio.

For the dairy cooperatives of Gujarat, the most significant development during the early years of Operation Flood was the creation of a federation to look after the marketing of the cooperatives' products.

During the 1960s, when Amul's ghee, cheese and milkspray were introduced, we had to depend on the services of Voltas and Spencers to handle the distribution, since we did not have our own distribution network in place. I was increasingly concerned that marketing was still not in the hands of the milk producers. Voltas and Spencers had, of course, done a wonderful job. But it was now time to replace them with the farmers' own marketing system. After all, my responsibility as an employee of farmers was to look after their interests and not the interests of Voltas and others.

Another concern that preoccupied me was that with more cooperative dairies coming up elsewhere in Gujarat, we could well be faced with a situation where milk cooperatives started competing against one another. I could not allow this to happen. In fact, in 1969, as Amul's General Manager I had signed an agreement with the neighbouring and large cooperative dairy at Mehsana, whereby they would produce butter and milk powder under the Amul brand name and we would distribute their products through our distributors. Thus, instead of competing with each other, we would combine forces and expand our markets while saving on advertising and brand building.

It had worked well with Mehsana and there was no reason why the same logic could not be extended to the remaining cooperatives in the state. Thus, in 1973, the Gujarat Cooperatives Milk Marketing Federation (GCMMF) was established. When I was made Chairman and Managing Director of the federation, I quit as the General Manager of Amul.

With the creation of GCMMF, we managed to eliminate competition between Gujarat's cooperatives while competing with the private sector as a combined, stronger force. We introduced what we called the 'balancing system' which meant that if there was excess milk in one dairy which they could not process, it would go to another sister cooperative dairy which had the processing capacity. It was a further extension of the collaborative, cooperative approach and it worked extremely well in Gujarat. So well, in fact, that today GCMMF is India's largest food products marketing organisation. This state level apex body of milk cooperatives has twelve district cooperative milk producers' unions, comprising 2.12 million producer members and 10,411 village milk societies with a daily average milk collection of some 4.5 million litres. The federation has ensured remunerative returns to the farmers while providing consumers with quality products under the two leading brand names of Amul and Sagar.

When our proposal for Operation Flood was being examined in the late 1960s, there was a serious debate about whether a country like India could – or rather, should – produce milk. Advanced dairying countries criticised us saying that here was a country which did not have enough food to eat and was importing foodgrains, whose population was growing at a phenomenal rate, and therefore, would have an increasing demand on foodgrains: in such a country, should valuable land be diverted from food production to feed production? In such a situation should we be producing milk, creating conflict between man and beast for land and its produce?

The second argument was that milk was not such a necessity, after all, in terms of nutritional value. Given all these facts, asked the advanced countries, could India afford animal products, including milk? Why then should we have a programme which would stimulate milk production? I thought the criticism did have some validity. But at the same time I felt that it was also indicative of the antagonism advanced dairying countries had to any move that a developing country makes towards self-sufficiency.

But the question did remain in my mind that any strategy we evolved must take into account the fact that with our agro-climatic conditions we could not give each cow or buffalo one acre of green grass, where all that the cow had to do was graze on it, morning, noon and night and produce 40 litres of milk a day. That type of dairying was obviously not for us. We simply could not afford to give every animal one acre of land when we were not giving every human being even one acre. We could not afford to keep the animals in a better shape and in better condition than our villagers.

That is why we ensured that the latter stages of Operation Flood included activities that centred on research into feed, animal health and animal nutrition. Our research helped in ensuring that our milk production should come from fodder and feed which man could not eat, which was produced not specially for our animals but was produced in the process of producing food for human beings. I think in the long run we proved to our detractors, and to the world at large, that even as technical, professional people we did look at dairy development not merely as the development of milk or animal husbandry, but in all its aspects. We were keenly aware at NDDB that every new project, each fresh activity undertaken, had to fit into a holistic approach to development.

TOUGH TIMES

We had anticipated that Phase-I of Operation Flood would be completed in five years. However, it took ten. Therefore, the cost, which was originally budgeted at Rs 96 crore, escalated to Rs 116 crore. We generated the additional funds required by increasing the selling price of the milk powder.

Most of the delay in implementing this phase of the project was caused in creating the Mother Dairies at Delhi, Calcutta, Bombay and Madras, which we needed to recombine the food aid commodities for supply to the four metro cities. Creating a market in these four cities was imperative for us in order to create seventeen Anands and to organise one million farmers.

One of the tools we decided to use for the clean and efficient marketing of milk supplied by the Mother Dairies was through bulk vending machines in the metros.

I had often watched what my wife did with the milk she bought. She did what every other Indian housewife did those days – she bought the bottled milk from the Amul dairy and boiled it. At Amul we had modern pasteurisers to heat the milk up to 72.4°C for sixteen seconds to kill all the harmful bacteria but never at temperatures higher than that, so that we did not destroy the good bacteria. But our housewives boiled the milk in their homes at 100°C. In effect, they undid everything we tried to do in our dairies!

The prevailing unhygienic conditions and practices make all housewives cautious in spite of the modern technology in dairying. It means that in India pasteurisation and packaging boils down to merely prolonging the shelf life of milk.

When we were planning for Operation Flood we decided to look at these problems afresh. My colleagues and I examined what our wives and mothers were doing. These were, no doubt, traditional habits. But it was these habits that had kept us healthy because 40 per cent of our cows and buffaloes have bugs and some of these filter into the milk. Boiling safeguarded the milk and also increased its life. But then why were we pasteurising it? Why were we bottling it if it was being unbottled? Why should we transport a pound of glass with every pound of milk and bring back that pound of glass, sterilise it and put an aluminium cap on it? Why could we not provide milk in bulk through machines?

If we could do this, we could also abolish the messy milk rationing system in our cities. So cash and carry seemed the best option. This concept, of course, created a furore in Delhi. All the pundits in the government, particularly in the dairy division, were convinced that bulk vending would never work. They said: 'You put a coin into our public telephones and you get neither the connection nor the coin back. If Kurien with his crazy schemes puts up machines for milk for the people of Delhi, and if these people get neither their coins back nor the milk, there will be riots in the capital; it will become a law and order situation and all kinds of calamities will befall – questions will be raised in the Parliament, the minister will lose his job ... .'

Our proposal for bulk vending machines was not approved. But I was persistent. I knew that bulk vending was the right choice. It would not only save a great deal of money, but it was an integral part of an efficient milk marketing system. I dug in my heels and I fought bitterly for three years to get the machines. My opponents were as determined and they stymied Operation Flood by not approving the machines. These machines were integral to our plan for the sale of milk through the Mother Dairy booths. Until I got the Mother Dairy booths working I would not be able to raise the money to replicate Anands in seventeen places in ten states of India.

I finally met Fakhruddin Ali Ahmed, the Minister who later became India's President. I developed a great liking and respect for him, even if he did not approve my bulk vending scheme. He said: 'Sorry Kurien, I cannot allow you to import these machines.'

'Sir, they only cost Rs 1.5 crore,' I said. 'That's not such a colossal sum for our country. Please allow me to import at least two hundred machines to begin with and then, subsequently, I will get them made here.' But he refused, saying that there was nothing he could do to help me.

'Are you then suggesting, Sir, that if I want to introduce bulk vending, I will have to design and build the machines myself?' I asked him.

'Yes,' he replied.

I was alarmed. 'Sir, you know that it cannot work like this,' I tried to plead. 'Is this not a deliberate attempt to make Kurien the fall guy – make him put up a scheme and fail so that everybody can then sit back and laugh?'

Fakhruddin Ali Ahmed remained unmoved. 'I'm sorry, Kurien,' he said. 'There's really nothing I can do. This is the only way out.'

Here was another major crisis facing me and, once again, I saw in it a challenge. We set ourselves a new task and we designed and manufactured the bulk vending machines ourselves in India. It took us four years out of a five-year project to design and get approval for these machines. This was the main cause of delay for the first phase of Operation Flood. I have taken the blame for it for a long time but I must now also put on record that the blame was not solely mine. The blame also lay with those who said bulk vending machines would not succeed and thereby delayed our project by four years.

A team of experts visited Mexico City to see milk-vending units in operation. The team approved the system and UNICEF provided a few units from the US. We needed to find someone who could make these machines workable in India.

During my brief apprenticeship with TISCO, I had a batchmate, John Prasad, who was a brilliant engineer. Being a bit rebellious, the company had dismissed him. I contacted John Prasad and told him about the vending machine and asked if he would be willing to modify the design to suit Indian conditions. He was very excited and extremely confident that he could do it. He took it up as a challenge. Even the making of the token for the machine was a trial of skills. It had to be made of an alloy which nobody could replicate. The token acceptance mechanism had to be so designed that it would accept only this token and no other. Minute details were examined, such as its diameter, thickness, weight and electro-magnetic properties.

We manufactured two hundred vending machines initially. The first Mother Dairy booth was built in Delhi. We approached the Director-General of the rehabilitation of the armed forces to give the job of the concessionaires to non-commissioned ex-officers of the army. After all, army officers are expected to be more disciplined than others! We said we would not give them salaries but would give them two paise per litre of milk sold. The more they sold the more money they could make. The concessionaires were selected and trained. I also demanded that space for the milk booths be given in residential areas, for the greater convenience of the consumer.

On the day the milk machines were to begin functioning in Delhi, there was tremendous apprehension. A survey among senior journalists revealed that most of them were sceptical about the success of such a sophisticated system; they believed that nothing which required meticulous maintenance could work for too long in Delhi. By and large, everyone was expecting all hell to break loose. But we were prepared to take a chance. I always believed that nothing is ever gained without undertaking some risk. There are so many opportunities that pass by and if we do not seize them, they are wasted. I found this opportunity to improve the marketing of milk for the dairy farmers and I grabbed it. Despite my critics and detractors, I had a commitment to the farmers and I had the conviction to push aside all obstructions and go ahead.

The bulk vending machines became a huge success and milk became available almost throughout the day at the mere push of a button. It was a great revolution in marketing of milk in India.

~

Ten years is generally considered to be a substantial period in the life of a project – even one that so boldly aspired to change the scene of the dairy industry in the country, as ours did. Since the food aid that we received was the largest ever in terms of money and volume, Operation Flood consistently sparked controversies, raised debates and commanded media and public attention. It was only natural, then, that at the end of its first decade, the progress of this massive project came under unprecedented scrutiny.

Ironically, most of the evaluations and appraisals of the project done by international agencies gave high marks to the programme. For instance, in 1981 a UN Inter-Agency Mission reviewed the achievements of Operation Flood-I and concluded:

> The project has shown that dairying in India is a powerful development tool, in that it can serve to provide nutritious high protein foods to city dwellers and at the same time provide income to the very poor people in rural villages ... .
>
> By creating a stable outlet for the milk produced in rural

areas, organised milk production in many cases doubled incomes in villages and thereby contributed to an improved standard of living.

The success of Operation Flood ... has also demonstrated that food aid can be used successfully for the development of local agricultural industries, given the right kind of institutional structure coupled with properly planned and integrated programmes for development ... Notwithstanding criticisms of Operation Flood-I from within and from outside India, which have not all been constructive and which seem at times to outvoice the praise, the scheme has justly earned from many quarters, the findings of the mission are favourable.

Strangely, however, recognition for the work from our own nation was not as forthcoming. Around 1982-83, NDDB – and I, personally – came under severe attack for Operation Flood. A rash of articles in the Indian media and the publication of a series of books, maliciously aimed at demolishing everything that we were trying to build, temporarily played havoc with the project's progress.

Curiously, the Institute of Social Studies at The Hague opposed Operation Flood strongly, publishing a number of books and papers in criticism. Two of them were written by Shanti George, a faculty member at the institute. In October 1983, the *Illustrated Weekly of India*, a magazine published by the Times of India group, carried a cover story by Claude Alvares titled, 'The White Lie'. It was a highly distorted and slanderous article, charging the NDDB's Operation Flood with being a complete failure and questioning the motives of the dairy board's leadership.

I responded to this cover story with a letter to the editor of the weekly, pointing out several distortions and asking, in all fairness, that the magazine provide equal space to present NDDB's side and put the programme in its proper perspective. The editor deemed it fit to publish a substantial part of this letter, but crucial parts were edited out to give the impression that the original article was not all that erroneous. The selective cutting of my letter gave the impression that we had sent a surly response that failed to address the criticisms. Similar articles appeared in other magazines, indicating a concerted campaign against Operation Flood. In any case, we prepared a detailed response, refuting all the allegations point by point, and later we printed this as a document for anyone who wished to have correct information about Operation Flood.

So incensed were our people at this smear campaign that many in NDDB wanted to sue the *Illustrated Weekly* in a court of law for libel and a plea was actually drafted. However, we settled for a complaint to the Press Council of India, which was eventually upheld.

The most charitable way to look at this sudden spurt in media criticism was that these authors believed – quite rightly – that food aid is bad. Indeed, all of us knew that food aid could sap a country's will to produce food; it could depress farm prices and could make the recipient increasingly dependent on aid until, at last, that country was destroyed. We have seen this happen in a number of places where 'charitable' donations have changed diets and undermined local agriculture, often to the point where these countries are eternally holding out their hands for more. But my colleagues and I were intensely aware of this potential trap and were determined that Operation Flood would not be a charity programme that would weaken India. We actually used the free aid during the initial stages to establish direct contact between farmers and consumers and to break the stranglehold of the contractors who had walked away with the profits in the past.

The other criticism levelled at us was that Operation Flood, in fact, transferred nutrition from the villages to the cities, thereby depriving the villagers in order to feed the urban dwellers.

It was certainly true that the poorer the farmer, the greater the temptation for him to sell all the milk and earn more money for other essentials. But there were other factors to be kept in mind here. Milk contains two important nutritional components – fat and protein. Milk fat is three times the price of vegetable fat, and even as the Chairman of India's dairy board I was unable to put forward the thesis that milk fat was superior to vegetable fat. It is not. However, the protein efficiency ratio of milk is slightly better than the vegetable protein. These arguments, however, hold no meaning for a starving man and it is unrealistic to say that this expensive food must be eaten by the poorest of our poor. Instead, I would say that he should keep a buffalo and sell the milk to those who can afford to buy it at a reasonable price, and spend one-third of the money he gets to recoup what he has lost and the other two-thirds to meet his needs.

The problem of nutrition in our country, as everyone knows, is not so much of quality but of calories. Our poor get only one meal a day. Malnutrition is the result of poverty and the cure for this can only be to increase the income of the poor. The labourers who build the skyscrapers in Bombay have no fancy houses to stay in. Is the answer, then, to tell them not to build skyscrapers but to go and build their own houses?

Similarly, the men and women who produce our milk, like those who produce rice or wheat, do not have enough to eat. But the answer does not lie in telling them to stop selling what they produce. According to me, the economically viable answer is that we should ensure a decent wage for the country's poor. The women who sell milk from their buffaloes should be paid a price for it, so that they can keep some of the money back for themselves and their children. Despite its critics, Operation Flood certainly increased the incomes of our farmers and we saw, over the years, that in many states, it doubled and tripled their income. As a consequence, as the farmers' income from milk grew, we found that, certainly, they and their families consumed more milk.

Our critics also accused us of increasing the country's dependence on imports of dairy equipment for the implementation of Operation Flood. It was true that Operation Flood did require substantial equipment for the new dairies. However, as was often the case, our critics failed to – or did not wish to – look at the larger emerging picture. While there were only fifteen manufacturers of dairy equipment in India before Operation Flood, by 1984 there were over one hundred and thirty manufacturers making dairy and allied equipment.

One of the outcomes of the spate of anti-Operation Flood publications was rather unexpected. I was informed that the Queen of Netherlands wished to visit Anand, meet me and find out the truth behind the torrent of accusations that was being thrown at NDDB and at me. She arrived, having read an impressive collection of the books and articles opposing and attacking Operation Flood.

When a head of state travels to our country the government generally attaches a minister to the visiting dignitary. In this case, as the dignitary was a woman, Margaret Alva escorted the Queen. I had never met Margaret Alva before. As our meeting commenced, she came and sat next to me on the Indian side of the table, as it were. The Queen of Netherlands began her cross-examination. Clearly, she had done her homework well and grilled me for a long time. At the end of it, when she was finally satisfied that she had left no questions unanswered, the Queen said, 'Dr Kurien, you have convinced me. Now I know what you are planning to do. My worries and the worries of the people in my country are certainly not justified and from now on we will be your strong supporters. In fact, Dr Kurien, when you next come to Europe, do come and visit us and I will arrange for you to come and speak at the Institute of Social Studies.'

When the Queen of Netherlands said this, Margaret Alva suddenly piped up and said, 'Your Majesty, whatever you might think of him, I think Dr Kurien is an MCP!'

I was astounded. This was the first time ever that we had met and she was accusing me of being a male chauvinist pig. I looked at her questioningly. Then she said, 'Please allow me to explain. Do you see the crest of the NDDB? It is a bull. It should, in fact, have been a cow. After all, NDDB is about dairy development. Doesn't this prove my point?'

I thought this was getting a bit out of hand and the time had come to say something in my defence so I looked at the honourable Minister and said, 'Madam, no bull, no milk.' The Queen burst out laughing. Margaret Alva was dumbfounded. She knew she could say nothing more. But after this incident we became good friends.

While the controversy surrounding Operation Flood raged around us, we also received encouraging words from a number of people. One such message came from rather unexpected quarters. In 1981, when Robert McNamara was retiring from the World Bank, I sent him a letter thanking him for the support that he had given us and mentioning that soon, I, too, would have to think of retirement. He wrote back acknowledging my letter and saying: 'While the Bank can get along without me, Operation Flood would never be the same without you. So I hope you will reconsider your retiring from it ... .' Such positive wishes always gave me the added impetus to go on with the struggle.

In Delhi, however, the fallout of the media criticism was by no means pleasant.

In November 1983, NDDB conducted a two-day seminar on the formulation of policies for breeding and nutrition of cattle in India. The first day of this seminar was held at Anand and the second, concluding day was held in Delhi's Vigyan Bhavan. Rao Birendra Singh, the then Agriculture Minister, was to deliver the valedictory address. It was a large and impressive gathering and to my shock and horror, in his address, the Minister began questioning the very basis of Operation Flood and the efficiency of its implementation. It was no secret that I had been having problems with Rao Birendra Singh but I never expected such a public display of his antagonism. I was left with no choice but to stand up and speak in defence of Operation Flood, after the Minister's address – which violated the normal procedure and protocol. Our confrontation was out in the open with this, and naturally, the media had a field day.

Close on the heels of this incident, on 2 December 1983, a Rajya Sabha member raised a question in the Parliament, referring to the allegations made in the *Illustrated Weekly* cover story. Rao Birendra Singh did not reply himself but in a written reply submitted to the Rajya Sabha, his deputy, Yogendra Makwana, stated: 'It has been decided to evaluate the working of the NDDB/IDC, as also the project implemented by them, by a few experts.'

When the report of this reply appeared in the press the next day, naturally, everyone at NDDB and IDC was extremely upset at this formal acceptance of criticisms that the ministry knew were unfounded. Many senior officers met R.P. Aneja, NDDB's Secretary, who brought some of them to me. They were agitated and wanted to know what should be done about the Minister's statement. I told them that if they were upset they should find a solution for it but it was for them to decide what to do.

Clearly, the officers discussed and debated this at length over a few days. Then, in an unexpected development, on 13 December 1983, Aneja brought to me a letter of resignation signed by over seven hundred officers of NDDB/IDC. In a dramatic and hitherto unprecedented move, the entire staff of a public-sector organisation had resigned in protest against their Minister's statement in Parliament. It was a moment of great distress for all of us at NDDB but also a time of challenge. The officers met it with tremendous strength. It was a moment that called for a resolute unity within the organisation and I was deeply touched that each and every member of the organisation stood firmly behind me like a rock. After this incident, we emerged far stronger and more inspired to fight those who wished to sabotage the project, and thereby the interests of India's dairy farmers.

Naturally, next day in the Parliament, Rao Birendra Singh had to face a barrage of questions on the mass resignations at NDDB/IDC and he quickly changed his tune. He said that he did not mean that an investigation would be carried out but that he would review the progress of Operation Flood.

While this controversy raged within and out of Parliament and the media, Prime Minister Indira Gandhi called and asked me what she should do to handle it.

'Madam, you must order an inquiry. There is nothing else you can do,' I told her. 'But please select such a person to head the inquiry so that no one can make any allegations about his findings.'

The Prime Minister first suggested that B.K. Nehru could head the inquiry committee but I told her that I knew B.K. Nehru personally and therefore, the investigation would perhaps not be as unbiased as it should be. She thought for a while and then suggested L.K. Jha, a distinguished civil servant from Bihar, former Governor of the Reserve Bank of India and Chairman, Economic Administration Reforms Commission, and former Indian Ambassador to the US. Since I did not know him personally, I agreed. All I had asked was that the person heading the committee be an outstanding person, whose integrity nobody would dare to question. In fact, he was reputed to be rather pro-private sector, too, and when such a person finally gave us a clean chit nobody would have the courage to raise their voice.

In February 1984, on the Prime Minister's instructions, the agriculture ministry constituted an evaluation committee under L.K. Jha's chairmanship. Vishnu Bhagwan, Joint Secretary in the

Ministry of Agriculture, Department of Agriculture and Cooperation, was appointed as the committee's Secretary. The committee also included four other distinguished experts in fields related to Operation Flood and its implementation. These four members were S.K. Rau, former Director-General of the National Institute of Rural Development; I.Z. Bhatty, Director-General of the National Council of Applied Economic Research; N.N. Dastur, former Director of the NDRI and P. Bhattacharya, former Commissioner of Animal Husbandry.

This high-powered committee was to evaluate the performance of NDDB/IDC with reference to the specific objectives of Operation Flood and assess its achievements. It would also, on conclusion of its evaluation, recommend measures to further streamline the implementation of the programme. The Jha Committee met over eleven months during which they investigated NDDB thoroughly. They visited all our offices, they met all our critics, including the most voluble ones, they visited the states where Operation Flood was implemented and held discussions with concerned ministers, officials and the milk producers. In addition to all this the committee also conducted an in-depth study of the project and its achievements against objectives, and the constraints faced during its implementation.

At the end of its term, in a detailed report presented to the government, the Jha Committee summed up its overall assessment in these words:

> By any standard, Operation Flood has been a successful programme, implemented with competence and dedication, for which the credit should go to the NDDB and IDC. Undoubtedly, the effort to promote the milk production enhancement programme needs to be stepped up a great deal but this does not detract from their overall performance.
>
> The effort has not been free from difficulties. The successful implementation of Operation Flood depended not only on the

efforts of the NDDB and IDC but also the agreement and support of the state governments concerned as well as the response of those engaged in the production and marketing of milk. The time taken in reaching agreements with the state governments as well as in organising the cooperatives differed from state to state and district to district so that the pace of implementation was not uniform but frequently lagged behind the ambitious targets originally adopted.

The Jha Committee was also unstinting in its commendation of both NDDB and IDC for the work done. The committee's report observed:

> While a number of people seemed very critical of the entire operation, the bulk of the evidence left us in no doubt that, overall, the NDDB and IDC have discharged their responsibilities with care and competence ... .
>
> As consultant, the NDDB has provided technical expertise of a high order ... The IDC and NDDB, between them, have some 800 professionals in diverse disciplines, of high quality and motivation. They have not only been able to set up, but also manage efficiently, dairies in larger cities ... while in some cities sick dairies have been handed over to them for management with successful results ... . Another area where the two organisations have distinguished themselves is in making innovations. Among these are the bulk vending system, the use of electronic fat testing machines, the extensive indigenisation of the dairy equipment manufacturing industry ... .
>
> High marks must also be given to IDC and NDDB for their success in management and marketing. The transportation of milk from surplus areas to deficit areas over hundreds of miles by road and rail, without being spoilt, pilfered or adulterated on the way has led to the emergence of a national milk grid in respect of a perishable product which traditionally could only cater to local consumption.

My colleagues and I could not have asked for a more impressive tribute. With such a positive evaluation from the Jha Committee, our critics were silenced. It was perhaps the best thing that could have happened to us at that time and it proved to be immensely useful for the dairy board. The committee's report confirmed Operation Flood's credibility within the country and encouraged the EEC and the World Bank, who were considering support for the project's third phase.

The report also made a number of extremely helpful recommendations – among them the need to merge the NDDB and IDC – which helped us grow much faster. In accordance with the committee's suggestion, NDDB was merged with IDC. This new body corporate retained the name of National Dairy Development Board.

The Government of India moved a Bill in the Parliament to declare the NDDB as an institution of national importance and to endow upon it corporate status by an Act of Parliament. Support came for this Bill across the board, irrespective of party affiliations. The Bill was passed in both the Lok Sabha and the Rajya Sabha in a record time of four days. This was a measure of expectations the Parliament had from this body. Thus on 12 October 1987, NDDB in its new avatar was created by an Act of India's Parliament – the NDDB Act 1987.

As a result of the Jha Committee's evaluation, the sanction and the implementation of Operation Flood-III was far swifter and smoother. But perhaps, most of all, this evaluation once and for all put to rest the raging media controversy.

It did not, however, put an end to my ongoing problems with Rao Birendra Singh. More than once, after a particularly difficult period when the Minister was obstructing our work at NDDB, I was forced to hand in my resignation, which was never accepted by the government. Things finally reached a point where I concluded that I did not want to stay if the Minister did not want me to stay. Coexistence in such cases is never possible.

There had been a time, more than a decade earlier, when problems with an earlier Minister – Jagjivan Ram – had led me to decide to resign. I was in Delhi to hand in my resignation, when I met an old friend, Vikram Sarabhai. He asked me what I was doing in Delhi and when I told him that I was there because I wanted to quit, he invited me to his room in the evening at the Ashoka Hotel. When I got there, I saw that D.P. Dhar and some other senior bureaucrats were also present. Sarabhai introduced me to them and told them what I intended to do. All of them persuaded me to think again, and promised me that they would 'take care of the Minister'.

Perhaps that earlier experience and the kind comments that her close advisers had made stood me in good stead this time. For when I went to see Indira Gandhi about my problems with Rao Birendra Singh, she picked up the phone and told the Minister in no uncertain terms: 'Don't do this. Leave Kurien alone.'

But Rao Birendra Singh refused to leave me alone and, perforce, I had to visit the Prime Minister once again and complain: 'Madam, this is the end. I'm not continuing any more. Since you spoke to him the last time, he has become even more malicious. I cannot carry on like this.'

Indira Gandhi looked at me and said very calmly, 'Dr Kurien, then I shall remove him.' And that is exactly what she did. She sacked the Agriculture Minister because he refused to stop interfering in the working of NDDB. Rao Birendra Singh came to know what was in store for him only three days before he was sacked. Immediately, he tried to extend a hand of friendship to me and cajole me to let bygones be bygones. But it was too late.

There are moments in life when it becomes imperative to protect one's stand. Most of the fights I have had with politicians and bureaucrats over the years have been essentially because I have been trying to hold on to my principles. I feel proud to say that I survived all those battles. The clash with the Comptroller and Auditor General (CAG) was an instance of this. It all began years before the man became CAG when, at a meeting, he said something that upset me. If I remember correctly, in the heat of the discussion, I said something quite rude. He was, understandably, upset.

Consequently, when CAG discovered that a firm of private chartered accountants conducted the NDDB audits, he found the excuse he was looking for to get his revenge. He told the then Minister of State for Dairy Development, Raghuvansh Prasad Singh – whose feathers also I had managed to ruffle – that as the Minister he should order an audit of NDDB. He explained to the Minister that although under the NDDB Act, the dairy board did not need to be audited by them since it was a government institution, under the Auditor General's Act we were obliged to allow the audit of the accounts of all government bodies.

The very next day the officers of the Accountant General of Gujarat arrived at the NDDB office in Anand demanding to see our books of accounts. I was away at that time so the NDDB officer who met him said, 'Dr Kurien is not here so we cannot give you the accounts.'

'I don't want Dr Kurien, I want the books,' declared the Accountant General. My colleagues at NDDB refused to show him the accounts in my absence. The episode did not end there. Arguments soon escalated and we received an official letter stating we were 'obstructing a government official from performing his legitimate duties. This is a serious offense and it can call for imprisonment ... .'

I came to a decision and informed my colleagues at NDDB that we would not show them the books of accounts because the motive was not correct. There were precedents in our system that when two government departments quarrelled they did not go to court because by doing so both departments would be spending government funds – public funds – to defend themselves. The normal procedure for such situations was that a secretary from the Cabinet Secretariat would try to arbitrate. Accordingly, the Secretary wrote to me to appear before him. I wrote back asking him to confirm that the CAG of India would also be appearing because any arbitration would be futile with just one party present. The Secretary was in a dilemma. He could not order the CAG to appear before him, as he ranks much higher than a Secretary. Once more the Secretary requested me to appear and once again I insisted that I would do so only if he assured me that both parties would be present. Finally he asked the CAG to send a representative. Two directors from the CAG's office came to the arbitration. The date was fixed and the Secretary began the process by asking me about my case.

'My case? What case?' I asked. 'I've been minding my own business. I want to know if they have any case.'

The directors said that they were doing their duty 'because the Constitution requires us to audit all government bodies' accounts.'

'If your auditing is so good then there should be no corruption in any government department,' I pointed out to them. 'Unfortunately we all know that corruption keeps increasing in government departments and I think it's probably because of some drawbacks in your auditing.'

They were upset and they asked me if I had anything to hide.

'Is it not true that you spend Rs 300 crore of public funds on your own department?' I asked them. 'I want to do an audit of that.'

They said I could not do so.

'Why?' I demanded. 'Why should you not allow me to audit? Do you have anything to hide?'

They told me I had no right to audit their department's accounts. 'Exactly,' I said. 'That is my reply to you, too. You have no right to audit my accounts.'

Friends within the ministry had informed me about all the details surrounding the Minister's order. I was therefore in a position to also point out to the directors that the order to audit NDDB's accounts had been put into motion without the due process being followed. I had some serious and valid objections and I voiced them without fear.

'You went to my Minister, told him some story and he in turn ordered the audit,' I said. 'You thought that by doing this you would win, because I would be under orders from my own Minister, who instead of protecting me is furthering your cause. You have forgotten one thing, however. My Minister has no right to make any request to the CAG. If he has any such request to make he has to first ask the Finance Minister who will then ask the CAG and only then can the order be acted upon. What is worse, the Minister ordered the audit half an hour before he demitted office and at a time when neither the Secretary nor the Joint Secretary were present. So this audit has been ordered without the due process being followed.'

The arguments continued in the same vein for a while. Finally the CAG's office took the matter to court. NDDB engaged an excellent lawyer, who argued that the entire order was illegal and also that the motive was mischievous. The case is pending in the court. Officers from the CAG's office had remained in Anand for six weeks hoping to get the order through. But we did not allow it.

The Act of Parliament that established NDDB states very clearly that the dairy board is autonomous. The point I wanted to make to the CAG and, indeed, to everyone concerned, was that if I was not prepared to fight for this autonomy – which in the first place was given to me by the Parliament – I would never be able to sustain that autonomy. In times of crisis, most people tend to surrender the autonomy that is given to them and this is very tragic. We at NDDB – and at Amul before that – always insisted on our autonomy, which we have never allowed anyone to violate. I have found, in over fifty years of confronting governments and defending cooperatives from political and bureaucratic interference, that when you begin demanding what is rightfully yours, there are many people even within the bureaucratic system who ensure that you retain those rights.

T.P. Singh, Secretary, the Ministry of Agriculture in 1972-75, was one such remarkable man. As Chairman of NDDB, which was a public society and trust promoted by the Ministry of Agriculture, I was obliged to go to Delhi once a month for a meeting with the Secretary of my ministry. During this meeting the Secretary would tell me all the things that I should have done and did not do and I would tell the Secretary all the things government should have done and did not do. It was a healthy exchange. At these meetings, generally the Joint Secretary and the Director too would be present.

At one particular meeting, after having completed our discussions, I got up to take leave of the Secretary, T.P. Singh, and the others when he told me that he had not finished. He asked me to sit down, pulled out a bulging file from his desk and showed it to me saying that he had only just come to know of its existence. It was a file on me, prepared by his colleagues in the ministry, particularly the Joint Secretary and the Director. These two bureaucrats, who were present in the room at that time, looked positively uncomfortable. In essence, the file said that it was illegal for me to be the Chairman of the dairy board and at the same time the Chairman of the IDC because it meant a conflict of interest. His colleagues, T.P. Singh informed me, sent this file to the law ministry, which concurred that my holding the chairmanship of these two bodies was, indeed, 'illegal'.

I stood up and told T.P. Singh that if this were the case, I would submit my resignation immediately from both NDDB as well as IDC. He looked at me and said, 'I have not yet finished, so please sit down. I do not know why my officers are doing such destructive work. We – this ministry – made Dr Kurien Chairman of the NDDB. This ministry then went ahead and made him Chairman of the IDC and now this same ministry starts a file saying this is "illegal". It is like a woman calling her own son a bastard. This is absolutely terrible.'

The Joint Secretary and the Director were visibly flustered and nervous at their Secretary's outburst. They began protesting meekly.

'There is no need to shed crocodile tears,' T.P. Singh said to them sternly. 'I know you fellows well. Now this file will remain here and only I will handle it. You are not to touch this file.'

He then told me that I could go but I should meet him again when I was next in Delhi. When I called on him during my subsequent visit to Delhi, he took me to the law ministry to meet the Secretary. He introduced me to the Secretary and then said to him, 'I sent you a file on Dr Kurien some time ago. Have you examined it? Have you read my note on it? '

The Secretary said he had.

'Is it "illegal" for him to be Chairman of the dairy board and the dairy corporation?' he asked.

The Secretary said it was not.

'Have you recorded your views on the file?' asked T.P. Singh and when the Secretary said he had done so, he asked to have the file back.

Taking the file, we returned to T.P. Singh's office. 'So Dr Kurien, we have won,' said this extraordinary bureaucrat. 'You see, what these fellows do not understand is that legal opinion is merely an opinion. It is not the law. Legal opinion depends on the briefing given and these chaps from our ministry gave the briefing. All I did was to give the Law Secretary my briefing. In my rather long career, Dr Kurien, I have been a legal officer, too, so I gave my legal opinion.'

Here was an officer I barely knew at a personal level. My interactions with him had been restricted to our monthly work meetings and yet he took the trouble to support me in the face of opposition from people within his own ministry, who for reasons best known to themselves, objected to the 'power' I wielded as Chairman of NDDB and IDC.

~

I have often been asked for my views on who could have incited and fuelled the anti-Operation Flood propaganda during the early 1980s, and why. I could never say with any certainty. But I was convinced that there had to be a multinational hand behind the materials published in Europe. I have constantly stressed that the real power of India lies in its people. Projects such as Operation Flood were meant to unleash this power. We Indians are an extremely intelligent people but we can progress as a nation only when we learn the secret of unleashing this positive power of the people. Whenever this happens it disturbs a lot of people – because they know that a giant is waking up.

As for the media criticism in India, certainly one of the people behind these virulent attacks was my erstwhile colleague, A.T. Dudani. We had been together in Bangalore's dairy research institute before I left for higher studies abroad. Since then Dudani kept in touch with me. I knew that he was unhappy at NDRI. He had hoped to be promoted to Director of NDRI and when this did not happen, he grew increasingly frustrated. When I became Chairman of NDDB, Dudani expected me to help him get a significant post somewhere. I did come to his help and, because I had respect for his ability, found him a position in the Delhi Mother Dairy. When it did not pan out, I placed him in IDC. When that, too, did not work out, I supported his return to the office of the Animal Husbandry Commissioner in the Government of India. There, instead of supporting Operation

Flood, he became a persistent critic, often going beyond the bounds of propriety in the way he treated confidential responsibilities. In the end, he was sacked by the ministry, an event for which he incorrectly blamed me. As a consequence, he held a lifelong grudge and made it his life's mission to attack me and my efforts.

I have often been accused of being unduly harsh towards critics. In fact, Operation Flood and I are indebted to some critics whose well-considered and sincere observations helped us to avoid mistakes and improve the programme. There are other critics, including Shanti George, who subsequently came to me personally and apologised, saying that I had been proved right in the final analysis and that they were wrong. I think they understood that I have never refused to listen to criticism that is honourable, criticism that is meant to improve, not defame. What we tried to do at NDDB was to create peaks of excellence. Maintaining excellence is always a painful job but we did achieve it in virtually all our projects. For us, excellence was particularly difficult to obtain when bureaucrats and politicians, who knew far less about the issues we were working on than we did, sat in judgement and continued interfering. I resented this intensely and in the process, I know I created a lot of resentment. But since this was to be expected, I also learnt to live with resentment and with criticism.

I am aware that I have also often been accused of being autocratic. Perhaps there were times in my career when I was, yet it was always as a response to a situation. If the situation demanded autocracy, I gave it autocracy. But never once did I let go of the larger vision and mission that had been ingrained in me by people like Tribhuvandas. The vision was that of democracy at the grass-roots and the mission one of placing the tools of development firmly in the hands of the dairy farmers. I told my critics that there was but one very simple test to see whether or not my concept of Operation Flood was flawed. All they had to do was to see whether the country had become more, or less, dependent on aid. I do not think there is space for doubt in anybody's mind on that score any more.

My NDDB colleagues and I were always intensely conscious that each of us, who had been given some responsibility, must, in whatever we did, keep the larger dimensions of total development in focus. It was our primary job to see that we brought the people into the development process. If my only intention had been to build Operation Flood without thinking of the larger development, I could have easily designed it in such a way that all these dairies were built and owned by the IDC and NDDB. After all, we had raised all the money. After all, the Government of India undertakings routinely build their factories all over the country and own them. But we did not do it because contrary to what many have often accused me of, neither the dairy board nor I was in the business of building our own empire. We were, from the very outset, in the business of building and expanding the empire of India's farmers. There is no other government agency in the entire country which, after spending more than Rs 1,000 crore, as the NDDB did, owns nothing. And in this, I take great pride.

STEP BY STEP

I WAS FORTUNATE THAT I ENJOYED THE SUPPORT OF ALL THE GOVERNMENTS that came to power. I have been – and continue to be – highly critical of our bureaucracy. Fortunately for us, within our bureaucracy there are a number of people – men and women – who are dedicated, patriotic and able. They are committed and principled people. I can never forget that many of these bureaucrats rallied to my support, going out of their way to give me a helping hand whenever I needed it. If I was in trouble it was always to these bureaucrats that I went. And they never failed me. This is a reality I cannot deny. We need the bureaucrats to look after the interests of our people. The tricky part has always been: how do we educate the bureaucracy to be truly public servants – servants of the people – rather than the bosses most of them continue to be?

As far as Operation Flood was concerned, once the prime ministers, the chief ministers and the ministers visited Anand and indicated that they wanted the project to take off, there were enough bureaucrats to support it and to neutralise those who opposed it. It was this combination of the approval of heads of governments and the support of bureaucrats in strategic positions, which further bolstered my courage to call a spade a spade and even to thumb my nose at some politicians, when the need arose. One of my truly memorable 'encounters' was with Jagjivan Ram when he was Union Minister for Agriculture and Irrigation in Prime Minister Indira Gandhi's cabinet in 1970.

I was at Krishi Bhavan in Delhi to attend a meeting with Prof. M.S. Swaminathan and others, to select a director for the CFTRI in Mysore, when I was told that Jagjivan Ram wanted to meet me urgently. I was about to tell the messenger that I would go after we had finished our meeting but Swaminathan astutely advised me: 'No. When the bada minister calls, you should go immediately. We will wait for you.'

I got up and left the meeting and was literally rushed to the Minister's chamber. Jagjivan Ram looked at me as I entered his room, asked me to sit down, smiled and said, 'Dr Kurien, I'm very happy to see you. I want you to help build a dairy in my constituency.'

I looked at him, a little confused, and asked, 'A cooperative dairy?'

'No, no,' said the Minister. 'A private dairy.'

I was stunned. I was not in the business of building private dairies. This was a totally unethical request and I responded rather violently, saying: 'It cannot be done.'

Jagjivan Ram could not believe what he heard and asked me to repeat what I had said. So I said it again, louder and clearer this time: 'I will not do it. I am here only to build cooperative dairies, not anybody's private dairy.'

'Is that so?' he asked.

'Yes,' I said rather angrily.

The Minister was obviously shocked at the way I spoke and he said curtly, 'You may go.'

That was the end of our meeting. I knew immediately that I had played it wrong. I should have agreed and said, 'Yes Sir, it's a very good idea. I'll think it over. I'll send you a note on it,' and thus played for time and then pointed out how NDDB's funds could not be used to build private-sector dairies. Instead

I had reacted very bluntly and, in hindsight, perhaps even very immaturely.

In all fairness, Jagjivan Ram was one of the ablest of our ministers. He was a seasoned politician and he was known to treat his officers very well – they all loved him. If I had handled his request more deftly, he would have been a great support to me. But I had been my outrageously outspoken self. Everybody warned me that I had made one of the biggest mistakes of my life in trying to tangle with him. They told me that Jagjivan Ram was such a skilled player that when he cut your throat you would not even realise it till your head rolled down. Certainly, I was in deep trouble.

Before too long, friends in Delhi reliably informed me that the Minister had asked for my file. Later, a sympathetic bureaucrat showed me that file and I saw that he had written there in red ink: 'Remove him'. I daresay the fact that he ordered my removal was received with great joy in the agriculture ministry. The justification given by Jagjivan Ram was that the Chairman of NDDB and the Chairman of IDC (those days they were separate bodies) should be two different people. He was a wily man and knew that it was IDC that had the money and, therefore, suggested that I should be removed from the chairmanship of IDC, not NDDB.

I realised that I had but one option left. I wrote a letter to Prime Minister Indira Gandhi, informing her about the situation and asked her for an appointment. I was called to Delhi. When I got there, V. Ramachandran from the Prime Minister's Office telephoned me and asked me if I could tell him what the problem was, so that he could brief the Prime Minister and that would help facilitate a decision. I told him what had happened and added that I felt it was not right on the Minister's part to terminate my services in this fashion.

Within two hours Ramchandran called me back to say, 'Sir,

the Prime Minister has just signed a letter and has sent it to the Agriculture Minister. The letter basically says, "Don't touch Kurien. Leave him alone." So your problem seems to be solved. If you still wish to meet her, I will arrange it.'

But I knew Indira Gandhi was a very busy person and since my work was done I did not disturb her. It was reassuring to know that she had supported me. Actually she was favourably disposed towards me from the word go. She had been to Anand along with her father and had stayed at my house. She saw the regard Jawaharlal Nehru had for me and for the work that was being done there. She was standing next to her father in Anand when he had come to inaugurate our dairy, and had heard him remark to me: 'I'm very happy that there are people like you in my country ... .' As far as Indira Gandhi was concerned, being appreciated by her father was enough. I had his stamp of approval and I knew that for this reason alone I could depend on her support in times of trouble. In turn, Rajiv Gandhi knew that his mother was very fond of me, and when he became Prime Minister, he too was equally supportive.

Although I was in a small town, I always had access to all our prime ministers. Initially it was largely due to Tribhuvandas Patel who was the Gujarat Pradesh Congress President, and Morarji Desai, who was extremely close to him and consequently to me and my family. There could have been no better backing for NDDB and the cooperatives than the ongoing approval of prime ministers. During my tenure as Chairman, I found myself often calling on their support for one reason or another.

~

In 1977, H.M.Patel became India's Finance Minister. Like Sardar Patel, he, too, traced his roots to this region – he came from Dharmaj village, just a few kilometres from Anand, and was an extremely close friend and supporter of Kaira's milk cooperatives.

He had seen the spectacular growth of the cooperative movement at Anand and was aware that because of what we had achieved at Anand, the country's production of milk had increased, leading to a complete halt of imports in milk.

H.M. Patel called me to Delhi for a meeting with other officials, where he shared with me his concern at the heavy outflow of foreign exchange due to rapidly growing imports of vegetable oil. He said that the country was already importing one million tons of vegetable oil worth Rs 1,000 crore and when he projected the consumption figures for five years ahead, he came up with an extremely worrying situation. It was his responsibility as the Finance Minister to see that precious foreign exchange was conserved and he asked me whether the Anand pattern of the cooperative structure could also be tried out in the area of vegetable oil, so that we could make the country self-sufficient in this commodity, too.

Contrary to what many have believed, it was not NDDB's idea to enter the vegetable oil market. We took on the project because the country's Finance Minister requested us to do so. Following the Anand pattern meant the procuring, processing and marketing of vegetable oil would have to be transferred to the hands of those who produced the oilseeds. The oilseed farmers would have to own and command the system and in the process eliminate the middlemen – the powerful and influential telia rajahs (oil kings) who interposed themselves between the producers of the seeds and the consumers of the oil. Only this could help us to bring to the producer a greater share of the consumer's rupee.

We evolved a Rs 700-crore project to restructure the production and marketing of edible oil and oilseeds and to cooperatise the vegetable oil industry. The project was launched in 1979. We planned to operate in seven states of India – Gujarat, Maharashtra, Andhra Pradesh, Karnataka, Tamil Nadu, Orissa and Rajasthan. By 1986 over 300,000 farmers had joined 2,500 oilseed growers' cooperatives across the seven states. The cooperatives were organised into unions and state federations which were responsible for the production, procurement, processing and marketing, as well as methods for improved yields.

Subsequently in 1989, the Government of India decided that there should be a comprehensive, integrated policy for the development of vegetable oil in India so that its imports, distribution and production would be coordinated. The import of edible oil was in the hands of the Ministry of Commerce; distribution with the Ministry of Civil Supplies; and production with the Ministry of Agriculture.

As often happens in government, there was a lack of coordination between these three ministries, with each following its own objectives, often contradicting the other two. This is why the Government of India evolved an integrated policy.

Market intervention was a central plank of the integrated policy for oilseeds and edible oils. In its wisdom, the government decided at the highest levels that NDDB should be that market intervention agency. Of course, within no time, NDDB – and I as its Chairman – came in for tremendous criticism for this from various quarters. By this time I had sufficiently hardened to criticism. I, and NDDB, could take any amount of abuse and criticism as long as we were confident that we were pursuing the right path, as long as we did so with integrity, and as long as we were allowed to perform.

The market intervention operation stipulated that NDDB should buy both oilseeds and oil, as it deemed appropriate, store these, mill the oilseeds and market the oil. As the market intervention agency it was our responsibility to bring stability to what had been an extremely volatile market for oil, maintaining the price above a certain limit to protect the farmer, and yet to keep it below a certain limit so that the consumer did not pay too much. In agricultural commodities, the trader – in this case the telia rajah – always made the bulk of the money, very often by manipulating information and the market. It is extremely difficult in agriculture to produce only as much as is required by the market. If the farmer produces 5 per cent more, then the trader who buys the produce beats down the farmer's price by as much as 50 per cent. If the production is 5 per cent less than the demand, the trader increases the consumer price by as much as 50 per cent.

Since our agricultural production is so largely dependent on the monsoons, it cannot be finely tuned and planned, and such variations are inevitable. In those days, the traders in the oil sector made tidy profits. They knew only too well that simply by buying and selling oil they made money; if they bought the oil, kept it for some time and then sold it, they made even more money; if they bought the oil, speculated and then sold it, they became millionaires. The ever-increasing greed of the traders exploited both the farmer and the consumer. And on us, at NDDB, the government placed the challenging task of achieving a stable price level that would be beneficial to both the farmer and the consumer.

By 1994, NDDB organised 5,348 Oilseed Growers' Cooperatives with around one million farmer members. We set up nineteen Oilseed Growers' Unions in the seven states and built several large, modern oil mills that were owned by the farmers.

An intrinsic part of the market intervention operation was to introduce into the market packaged edible oil sourced from the oil cooperatives. By the mid-seventies, we were clear that the biggest and the most powerful friend of the milk producers of Gujarat was the trade name of 'Amul' which stood for quality, reasonable price, and trust. The biggest assets of the milk producers of Gujarat were the distribution and marketing channels that we had set up nationally for Amul products.

Realising the importance of a trade name we introduced 'Dhara' which means 'stream', as the trade name for our packaged oil and strengthened the position of the oilseeds cooperatives.

Soon Dhara, too, would stand for quality, reliability, freedom from adulteration and reasonable price. There was only one difference – unlike in the case of Amul where I had twenty-five years to achieve success, for Dhara I had only a few years. Therefore, to ensure for the product a relatively 'instant' success, I introduced Dhara as the lowest-price branded oil in the market, but with a quality equal or better than the high-priced brands.

Since GCMMF already had a network of over 500,000 retail outlets to sell Amul products all over India, the marketing of Dhara was given to GCMMF. Also, nobody else would have agreed to market Dhara at an incredibly low margin of only 1.5 per cent!

Dhara was introduced specifically to give consumers oil at a reasonable price. In the beginning, as part of our marketing strategy – to stimulate marketing and ensure consumer acceptability – we put Dhara at a lower price than the cost price. We could do this because the Government of India had agreed that they would give NDDB 150,000 tons of imported oil for its market intervention operations. The government also clearly stated in writing, and with Cabinet approval, that it would meet any losses on market intervention operations that NDDB might incur. During the initial two years we incurred no losses because we were also being given imported oil, and by marketing this we made a profit, which more than offset the loss on Dhara. Subsequently, however, government was not able to give us any imported oil because there was a foreign-exchange crunch.

If a shortage of an agricultural commodity is to be removed, it can be done if the farmer who produces it is paid a price that gives him some profit in producing it. With this in focus, we brought the price up to a level that was not only satisfactory from the consumers' point of view, but provided a real incentive to the producer. Very rapidly Dhara established a reputation for quality and purity. The ultimate motive for our market intervention operations was to ensure that the product from the producer organisations reached the consumers directly. Oil was being traded at the oil exchange between traders; we wanted to bypass the oil traders and the only way to do it was by gaining a consumer mass market. We managed this to a certain degree, despite initial losses.

It is true that we did not see as astounding a success in the edible oil sector as we had in the milk sector, but that was also because of the difference in the two commodities. While unscrupulous traders could hoard, stock and hold on to oil they could not do so with a highly perishable commodity like milk. Even so, the powerful oil lobby was definitely shaken by Dhara. Although Dhara was a genuine consumer marketing success, becoming the leading branded oil within a matter of one to two years, it seems not to have been appreciated because it was not achieved by a multinational. The tragedy of India is that we frequently have no respect for Indians, for Indian efforts and for Indian successes.

This reminds me of a remark made to me, at the height of Amul's success, by a high-ranking official of the Government of India in Delhi. He said, 'Kurien, these chaps at Nestle are truly great people. What marvellous things they have achieved. You should go and see how well they run their dairy.'

'And shall I tell you how much better the British ran this country?' I retorted. 'So then, shall we call the British back? '

He had nothing further to say to me.

If we look at vegetable oil before NDDB was asked to get into the act and after it did so, India's edible oil import bill came down drastically, from Rs 1,000 crore to Rs 165 crore, saving precious foreign exchange. Dhara did not do too badly and its sales quickly touched Rs 1,000 crore per annum. When the branded oil market was growing by 12 per cent annually, Dhara was growing at the rate of 35 per cent.

Whether one liked it or not, for years, oil was considered an essential commodity and this created a string of complications. Added to this was the fact that there was more than one ministry involved in the edible oil sector. The Ministry of Commerce with its State Trading Corporation (STC) had a problem if there were no imports. The civil supplies ministry was supposed to help consumers by selling them imported oil at a low price. So if the country became self-sufficient in edible oil, STC could not import and civil supplies would have to base its public distribution on Indian oil, not imported, which meant that consumers would have got it at Indian oil prices. However, when the government was bound to sell it at a lower price through the public distribution scheme, how would they subsidise it? The finance ministry refused to do so, and therefore, what would happen to public distribution in oil? All these issues were the obstacles in our path.

Despite such a complicated scenario, however, I could not agree to a structure where a businessman and speculator bought the oil and became a millionaire. Such people, under the private sector and in the name of liberalisation, could not be allowed to have a free reign to exploit the farmers. That would only indicate that the government was abdicating its right and responsibility to govern and this was the reason, I suspect, NDDB was asked to try its hand at cooperatising vegetable oil. Our strategy was very simple: turn the farmers loose – give them adequate incentive – for they are the only ones who can produce more.

The fact remains that in about five years the country, by 1993, became virtually self-sufficient in oil. The oil lobbies of other countries were unhappy because India's oil imports dropped from more than 2 million tons to under 200,000 tons. Within the country, I think we did manage to bring about some discipline. We took care of the telia rajahs but we did pay a heavy price. Many NDDB officers were physically attacked. On seven occasions our cooperative oil mill in Bhavnagar was set on fire. But we were not deterred from the task we had set for ourselves.

The scoundrels, who for decades exploited both the farmers on the one hand and the consumers on the other, were waiting eagerly for our edible oil project to fail. They said that unlike milk, oil was a very slippery business and Kurien was bound to slip and fall flat on his face. Ultimately, they were the ones who tumbled and fell. I am sure I made lots of people very unhappy, but for me it was more important that I made hundreds of thousands of oilseed farmers benefit.

~

During the time when NDDB's market intervention operations in vegetable oil were on, my wife and I happened to be in Delhi on a personal visit. One evening, a sympathetic bureaucrat called me in a bit of panic and said, 'Sir, all hell is breaking loose and you better do something about it very fast. The Civil Supplies Minister is scheduled to meet Prime Minister Rajiv Gandhi today and you must act before he does. We have some 27,000 tons of vegetable oil in stock and we wanted to give it to NDDB for the vegetable oil project. We have just learnt that the Minister wants to sell it at half the price to a private party and pocket part of the money. So Sir, please do something and save the situation. He will then want to import vegetable oil because there's a lot of money to be made on imports.

I asked the bureaucrat at what time the Minister was meeting Rajiv Gandhi, the then Prime Minister, and he informed me that the meeting was scheduled for seven p.m. I looked at my watch. It was five p.m.

'Look at the time,' I said to him. 'You're telling me now. How the hell do you expect me to get an appointment with the Prime Minister within two hours?'

Obviously some rapid damage control was needed and I

telephoned V. George, an officer at the PMO, and told him I had to meet the Prime Minister urgently before seven p.m. George said he would try his best. A little while later he called me to say that I could meet the Prime Minister at six forty-five p.m. I thanked George for his help and rushed off to meet Rajiv Gandhi.

Rajiv Gandhi greeted me and said, 'Yes, Kurienji, what is it that you wanted to see me about? Is there a problem?'

I told him what I knew – that his Minister was sitting on a stock of 27,000 tons of vegetable oil and that in the next fifteen minutes he would come to meet the Prime Minister to tell him that he wanted to sell it at half price because it had 'gone bad'. I told Rajiv Gandhi that NDDB was willing to buy the entire stock at full price and that although NDDB – a government outfit – needed that oil desperately, the Minister did not want to give it to us.

'Are you sure?' asked Rajiv Gandhi. 'You know all this for a fact?'

'Yes,' I replied. 'And he's coming here to slip it in in the course of discussing other matters with you.'

'Is that so?' the Prime Minister asked.

I assured him that this would happen and then, having given him all this information, I asked him if I could take his leave.

'No,' said the Prime Minister. 'You will sit down here and participate in this meeting.'

I was rather reluctant to do so and said, 'But I have not been invited to this meeting, Sir.'

'Well, I'm inviting you. You will sit here, Kurienji, and take part in this meeting,' said Rajiv Gandhi.

Soon the Minister, the Secretary Civil Supplies and three other officers arrived. I stood up when they walked in. The Minister glanced at me and Rajiv Gandhi explained that since I was interested in vegetable oil and since the market intervention operations of NDDB were on, he thought I should be present at the meeting too. The then Cabinet Secretary, T.N. Seshan, was also present. We sat down and the meeting began.

Predictably, the Minister first talked about a couple of other issues and then told the Prime Minister that there was 27,000 tons of oil in stock which was going bad and that he wanted to sell it for half the price, since it was better to make at least half the money rather than wait and allow it to be a total loss. Clearly he was unaware that I had been tipped off about this. Rajiv Gandhi looked at me questioningly. I said, 'Sir, I am prepared to buy the oil at whatever price.'

'Are you are prepared to pay the full price?' Rajiv Gandhi asked. I assured him that NDDB would pay the full price for the oil.

'I don't understand this,' said the Prime Minister, turning to the Minister. 'Here is the public sector willing to pay the full price. You don't want to give it to them but you want to sell it at half the price. What sort of policy is this?'

Seshan, the incorrigible, irreverent fellow that he was, suddenly piped up: 'Sir, this is called the dog-in-the-manger policy!'

Rajiv Gandhi looked at the Secretary and said: 'Give all the oil to him and it should be given to him within twenty-four hours. Okay, what's the next item on the agenda?'

But I still had to say my bit to this gathering so I continued, 'Sir, I want to tell you something about this Minister. The telia rajahs in Bombay are shouting and screaming that the Honourable Minister has made a crore of rupees during the last import. If crooks like the telia rajahs are broadcasting this information about your Minister in the streets of Bombay, what is your government's reputation worth?' I then asked him once again if I had his permission to leave. The message went home.

As I walked out of the Prime Minister's room, Seshan stuck out his hand to shake mine and later remarked to me that he never knew I was capable of saying such things. The Minister has never forgiven me for my role in it. Even today whenever I accidentally run into him somewhere, he turns his face and looks away.

Rajiv Gandhi was a good man, an uncomplicated person. The only mistake he committed was to allow all sorts of corrupt ministers to surround him. When he visited Anand to inaugurate the Ravi J. Matthai Library at the Institute of Rural Management, Anand, he shared with me a problem he faced.

'Kurienji, I made a terrible mistake,' he said. 'I thought I must help my constituency, Rae Bareilly, so I caught hold of an industrialist and persuaded him to spend some Rs 100 crore to start an industry there. But I realised that after spending all that money, 2,000 jobs were created out of which some 1,980 jobs went to people outside my constituency. Since my people were not too educated, they got the jobs of sweepers and I became even more unpopular. After coming here I realise what I should have done, so Kurienji, would you set up an Anand there?'

I explained to him it was not that easy to 'set up an Anand'. The entire process had to be streamlined – first, milk cooperatives must be formed and then milk must be produced, so some cattle-breeding had to begin. Only when milk production was adequate could a dairy be built and all this took time. It would have to be a long-term proposal. He agreed and we began the work to set up a dairy cooperative in Rae Bareilly when, very tragically, he was assassinated. Much later, I called Sonia Gandhi and told her that she should take a look at the project her husband wanted to launch in his constituency. She said she would come but that we should take someone else to actually look into the details of the project. We decided to ask Balram Jhakar, who had served as Minister for Agriculture in Rajiv Gandhi's government, to accompany us. When we got to the airport we saw that Sonia Gandhi's daughter Priyanka accompanied her. Priyanka explained to us that since her mother was visiting Rae Bareilly for the first time after her father's death, she thought she should be there to hold her hand and give her moral support.

At Rae Bareilly we saw what magic the Gandhi name weaves. At least a hundred thousand people were there to see Sonia Gandhi and hear her address them and very soon they started calling out: 'Priyanka! Priyanka!' It was quite amazing. I asked Sonia if she would like to go across and meet some of these people and she said she would but there were barricades. I assured her that barriers could always be broken. I took her towards the people and we removed some of the barricades. She walked into the women's enclosure and the women started weeping and calling out to both of them. I think she must have felt at that time that, with this kind of an emotional response attached to the Gandhi name, she could not afford to turn her back and walk away for too long.

The difference between Rajiv Gandhi's request to me for a favour and Jagjivan Ram's request way back in 1970 was that Jagjivan Ram had asked for a private dairy. His project had no element of helping the poor. I could never afford to lose sight of the fact that all of NDDB's projects necessarily had but one central mission – that of empowering the poor.

~

Much before NDDB began functioning, Amul had already demonstrated what small and marginal farmers could achieve as empowered members of cooperatives. When farmers join together, begin to cooperate, they are a power to be reckoned with. Therefore, ministers, too, had to be very careful before they criticised the Amul dairy. If they attacked Amul, they attacked these hundreds of thousands of farmers and their families; if they damaged the Amul dairy, they damaged the farmers of Kaira district. Much of the strength of the cooperatives derived from their democratic character and this was effectively displayed to another VIP visitor to Anand.

There was a time in 1990 when I ran into some rough weather with the Government of Gujarat. The dairy farmers of the Kaira Cooperative invited Devi Lal, who was then the Deputy Prime Minister as well as Agriculture Minister, and Chimanbhai Patel, Gujarat's Chief Minister, to Anand for a meeting. When the two politicians arrived they found around fifty thousand farmers gathered there. Devi Lal walked into their midst and started chatting with them. He asked one of the farmer leaders, 'Which party do you belong to?' and the farmer answered 'Janata Dal'. Devi Lal smiled. Then he went to the next one and asked him the same question, expecting the same reply, but this farmer answered that he belonged to the Congress Party. Devi Lal appeared a little shaken. He then went to a third leader with the same question and was told that he belonged to the Bharatiya Janata Party.

By this time the Deputy Prime Minister was thoroughly confused. He asked them how it was possible that although they had such different political affiliations, they came together as one group here. And the farmers and farmer leaders explained to the politician that within the cooperative, they were not party people, they were, first and foremost, farmers. In the course of the discussions that followed, what these fifty thousand people sitting there told the two ministers by implication was: 'If you attack Dr Kurien you attack a man who is working for us. Don't you know better than to do that?' From that moment onwards, Devi Lal became my ardent supporter and Gujarat's Chief Minister immediately released a statement saying that he supported me entirely and that I was 'a great Gujarati'. For me, nothing could have been more rewarding than the faith that, once more, Kaira district's farmers placed in me that day.

At many times in my life, I have been asked why, with such massive support behind me, I have never stood for elections. My answer has always been the same. If I ran for office, I would completely lose my value. To unite our farmers irrespective of caste, religion or political affiliations, for the sake of economic survival, is of the greatest importance to India. This is the way to unite the nation. Our politicians, unfortunately, have honed to a fine art their skill in dividing the nation into Hindus, Muslims, Sikhs and Christians, into Kshatriyas and Patels and Jats, into Gujaratis and Bengalis, into Hindi-speaking and non-Hindi speaking people. It is my belief that only the success of economic movements like the one spearheaded by Kaira Union will overcome these artificial and destructive divisions in our nation.

Another question often posed to me over the years is what I would do if I were ever made Union Agriculture Minister. I think my first step would be to step down as Chairman of the Indian Council of Agricultural Research (ICAR), of which the Minister is the automatic Chairman. Then I would reduce the number of ICAR members from ninety to twelve, for having ninety members is virtually trying to run a circus. I would ensure that the researchers got not more than 50 per cent of their research funds from government coffers and the rest would come from the industry. This way they would be answerable to the industry and would also get rewards for their good work from the beneficiaries. And lastly, I would trim the bureaucracy and instead promote and encourage independent, democratic structures which people themselves would command. In short I would ensure that Krishi Bhavan is a true Krishi Bhavan and not a 'Kursi' Bhavan as it has become.

Our bureaucracy today is too bloated and therefore it is burdensome. For example, 95 per cent of the agriculture budget goes into paying the staff's salaries and I would not be surprised if the remaining 5 per cent goes towards the maintenance of its jeeps. Where is the planning in that? As an interested and concerned citizen who has witnessed our planning process for the last five decades, I can see why the fruits of development today are not commensurate with the money spent.

In many ways, Rajiv Gandhi's famous statement, about only fifteen paise reaching the bottom when hundred paise are released from the top, said it all. The solution can only lie in creating democratic structures which people themselves command, instead of the bureaucracy. The place for the IAS officer is in the secretariat and not in the field. The IAS officer is basically an aya-ram gaya-ram. He is transferred at regular intervals and it is almost impossible for him to show commitment when he knows he is going to be transferred in a short span of time. I have never understood how, for instance, the Agriculture Secretary can be a person who does not know agriculture. Somebody who passed some competitive examination thirty-five years ago, is today suddenly placed in this post, when until yesterday he was, perhaps, the Law Secretary, and the day before that he was the Defence Secretary. What a strange system this is of administering the country. I am convinced that the IAS, in its present form, will have to be abolished sooner or later. There is no other solution.

If we depend – as we have too long depended – on bureaucracies and politicians and not on our people to deliver the goods, then there is very little that we will achieve as a nation. The bureaucrats and politicians will only become stronger. It will do well to keep in mind that the principal enemy of Anand was always the Milk Commissioner and the milk departments of government – far more than any Polson or private traders.

~

In 1967, as Chairman of NDDB, I was asked to be a member of a high-powered committee, set up by the Government of India, to look into cow protection. It was a collection of rather individualistic and interesting personages. Justice Sarkar, Chief Justice of the Supreme Court, was appointed its Chairman. Among the other members of this committee were Ashok Mitra, who was then Chairman of the Agricultural Prices Commission, the Shankaracharya of Puri, H.A.B. Parpia, Director of the Central Food Technological Research Institute in Mysore and M.S. Golwalkar 'Guruji', the head of the Rashtriya Swayamsevak Sangh (RSS), the organisation which had launched the entire cow protection movement.

When this committee was set up one of the first questions to be raised in the Parliament was: 'Is there any Muslim on this committee?' The reply given was: 'Yes, there is. Dr Parpia.' Whereupon Parpia got furious and wrote to the government saying: 'If I have been put on this committee because I am a Muslim, I herewith submit my resignation.' He told the government that since he was not a practicing Muslim, they better think again and put some mullah on the committee! However, the government managed to pacify him and he stayed on.

For some inexplicable reason, the Shankaracharya and I took a spontaneous and mutual dislike to one another. I still recall my first meeting with him. He strode into the room, bare-chested, carrying an ankush (trident) in one hand and a rolled up deerskin tucked under his other arm. He walked up to the chair next to me, spread out his deerskin on the seat and sat down. In those days I used to be a heavy smoker and I thought to myself that if he did not need permission to carry a deerskin, I did not need permission to smoke and I continued smoking. Unfortunately, each time I took a puff and exhaled the smoke, it would move in his direction. The Shankaracharya glared at me, made some angry noises, snatched up his ankush and deerskin and moved down a few chairs away from me. He continued to glower at me from his new position and I continued to smoke. Justice Sarkar, who was watching this little sideshow delightedly, leaned forward, tapped me on the shoulder and said, 'Dr Kurien, may I have a cigarette too?' That was Justice Sarkar. A great man.

Incredible as it might seem, this committee met regularly for twelve years. We interviewed scores of experts from all fields to get opinions of all shades on cow slaughter. It was a tedious and time consuming process. My brief was to prevent any ban on cow slaughter. It was important for us in the dairy business to keep weeding out the unhealthy cows so that available resources could be utilised for healthy and productive cattle. I was prepared to go as far as to allow that no useful cow should be killed. This was the point on which the Shankaracharya and I invariably locked horns and got into heated arguments. I constantly asked him, 'Your Holiness, are you going to take all the useless cows which are not producing anything and look after them and feed them till they die? You know that cannot work.' He never had any answer to my query.

For twelve years the Government of India paid the committee members to travel to Delhi and attend the meetings. We continued like this and it was only when Morarji Desai became Prime Minister that I received a little slip of paper, which said, 'The cow protection committee is hereby abolished.' We were never even asked to submit a report.

However, one rather unusual and unexpected development during our regular committee meetings was that during that time, Golwalkar and I became close friends. People were absolutely amazed to see that we had become so close that whenever he saw me walk into the room he would rush to embrace me. He would take me aside and try to pacify me after our meetings, 'Why do you keep losing your temper with the Shankaracharya? I agree with you about him. But don't let the man rile you. Just ignore him.'

Golwalkar was a very small man – barely five feet – but when he got angry fire spewed out of his eyes. What impressed me most about him was that he was an intensely patriotic Indian. You could argue that he was going about preaching his brand of nationalism in a totally wrong way but nobody could question his sincerity. One day after one of our meetings when he had argued passionately for banning cow slaughter, he came to me and asked, 'Kurien, shall I tell you why I'm making an issue of this cow slaughter business?'

I said to him, 'Yes, please explain to me because otherwise you are a very intelligent man. Why are you doing this?'

'I started a petition to ban cow slaughter actually to embarrass the government,' he began explaining to me in private. 'I decided to collect a million signatures for this to submit to the Rashtrapati. In connection with this work I travelled across the country to see how the campaign was progressing. My travels once took me to a village in Uttar Pradesh. There I saw in one house, a woman, who having fed and sent off her husband to work and her two children to school, took this petition and went from house to house to collect signatures in that blazing summer sun. I wondered to myself why this woman should take such pains. She was not crazy to be doing this. This is when I realised that the woman was actually doing it for her cow, which was her bread and butter, and I realised how much potential the cow has.

'Look at what our country has become. What is good is foreign; what is bad is Indian. Who is a good Indian? It's the fellow who wears a suit and a tie and puts on a hat. Who is a bad Indian? The fellow who wears a dhoti. If this nation does not take pride in what it is and merely imitates other nations, how can it amount to anything? Then I saw that the cow has potential to unify the country – she symbolises the culture of Bharat. So I tell you what, Kurien, you agree with me to ban cow slaughter on this committee and I promise you, five years from that date, I will have united the country. What I'm trying to tell you is that I'm not a fool, I'm not a fanatic. I'm just cold-blooded about this. I want to use the cow to bring out our Indianness. So please cooperate with me on this.'

Of course, neither did I concur with him on this nor did I support his argument for banning cow slaughter on the committee. However, I was convinced that in his own way he was trying to instil a pride across our country about our being Indian. This side of his personality greatly appealed to me. That was the Golwalkar I knew. They had accused him of plotting the murder of Mahatma Gandhi but somehow I could never believe it. To me he came across as an honest and outspoken man and I always thought that if he were the Hindu fanatic that he was made out to be, he would never have been my friend.

During the last days of his life, when he was ailing and in Poona, he called all the state heads of the RSS to talk to them because he knew he was dying. After Golwalkar passed away, one of these men turned up at my office to meet me. 'Sir, I am the head of the RSS in Gujarat,' he said, introducing himself. 'You probably know that Guruji is no more. He had called all of us to Poona and when I told him that I was from Gujarat, he said to me: "When you go back to Gujarat, please go to Anand and specially convey my blessings to Dr Kurien." I have come to convey this message to you.'

I was deeply touched and I thanked him, expecting him to leave after passing on the message. But he continued: 'Sir, I want to ask you a question if you don't mind. You are a Christian. Of all the people in Gujarat, why did Guruji send his blessings to you and you alone?'

I asked him why he did not ask Guruji this. I could give him no answer because I really did not know.

FROM ORGANISATION TO INSTITUTION

As OPERATION FLOOD TOOK CONCRETE SHAPE AND NEWS OF THE achievements of India's dairy cooperatives spread across the world, a number of foreign dignitaries expressed their desire to visit Anand to see for themselves the miracle that dairy farmers had created. A string of VIPs arrived one after the other including Britain's Prince Charles, HRH Queen Beatrix of the Netherlands and the British Prime Minister, Lord James Callaghan.

In March 1979, we had a very distinguished visitor – the Soviet Premier Alexei Kosygin. He was taken around on a guided tour of the villages, the dairy complex and the plant. After the tour he met me and said: 'Anyone who has seen the dairy plant, anyone who has looked into the eyes of the farmers in your villages here, knows what you are doing. You are eliminating the bloodsuckers. You are giving strength to the people. You are constructing structures and institutions of the people. What you are doing is absolutely correct.

'But Dr Kurien, you took thirty years to achieve this in milk – an entire lifetime – and you faced a lot of problems. When you set up the milk cooperatives nobody really realised what you were up to. If you try to do the same for vegetable oil, where big people are involved, everyone will know. Tomorrow, Dr Kurien, you may think of doing the same for cotton or jute and you will then upset the biggest people in India. Besides, you will take thirty years to do it in vegetable oil and then another thirty to do the same for cotton and jute. Such significant social and economic changes that you are trying to bring about, should not be done in this leisurely fashion. These changes should be brought about quickly – suddenly – in all directions, all at once. It has to be a revolution. For if you do it slowly, Dr Kurien, you will be shot down like a dog.'

After issuing such an ominous warning to me he then invited me to (erstwhile) USSR to see the changes the revolution had brought in. I was instructed by the Government of India to accept Kosygin's invitation. I went to the Soviet Union, along with some colleagues and it was one of the most miserable experiences I have had. From the moment we landed in Moscow we had the KGB swarming all around us. We were steered into a waiting bus. Every move was monitored and each activity on the itinerary was completely regimented to the last detail. We were expected to fall in line without a murmur.

I began protesting, knowing that the KGB would report every remark of mine. By that evening I was completely fed up of being herded around like cattle and I decided that nobody was going to take us for granted just because we were Indians. Therefore, when one of their women officers came to say that I was expected to report downstairs for dinner in a short while, I simply lost my cool. I told her that I would not go anywhere at any fixed time and that I first wanted a drink. I demanded a bottle of whisky and I think I just revolted so loudly that the KGB who were in attendance rushed off to inform their seniors that 'the foreign guest was out of control'. Two ministers were sent to my room to apologise to me. I took this chance to also tell them that I refused to travel by bus any more and demanded a car. I got that as well.

I was also completely unimpressed by their dairy industry. It was a total mess. The cattle were in bad shape. The cows were owned collectively and since they belonged to the state, it meant that nobody owned them and therefore no one took care of them.

Similarly, the dairies, too, were owned by the state and their condition was dismal. I immediately saw that this was not a system that would work efficiently and I returned to India even more convinced that, quite contrary to what Kosygin had advised me, change must take time. Basic social and economic change needs to be brought about gradually and the more carefully and thoughtfully it is effected, the more permanent it will be.

Evidently we were doing something very right in our country because over the years many countries across Asia, Africa and Latin America expressed interest in adopting the Anand pattern, sending officials and farmers to learn from our experience. There can be no higher praise than this. One of the countries, which showed an interest in replicating the Anand pattern of cooperatives in milk, was Pakistan.

In 1982, the Pakistan government wanted me to come to their country and help them set up dairy cooperatives but at that time the relationship between the two countries was so fractured that they could not make a direct request for me. What they did, instead, was to contact the World Bank and ask them to send a mission to Pakistan on the condition that 'Kurien of India would lead it'. I was asked to lead the mission to Pakistan for a period of five weeks. Though I could not afford to spend five weeks away from my work, I agreed to be in Pakistan during the first week, position my team, discuss what they should look into and then go back again during the last week to wrap up the project and meet their ministers. Hardison was the senior officer from the World Bank on this mission and the main worry for him and his colleagues was how to 'protect' me from the Pakistanis. To their astonishment they found that every evening one Pakistani minister or the other invited me out for dinner. They did not understand it because we were supposed to be 'enemies'.

Naturally, the first minister I met was Pakistan's Minister of Agriculture. He noticed the symbol of NDDB – a stylised version of the Mohenjodaro bull. The Minister chided me gently, 'Dr Kurien, I'd like you to remember that Mohenjodaro is in Pakistan, not in India, so what you have here is a Pakistani bull. You are welcome to it, though. You can keep it. But since you have stolen our bull you owe us something. You must help Pakistan in its dairy development.'

During the first week I also met the Finance Minister of Pakistan. He, too, was extremely enthused about a dairy development programme for the country. 'Look at India today,' he said. 'In fifteen years you have completely stopped the import of milk and all milk products. Everything is being produced in India. My country had a better start than you but you have stopped imports whereas in Pakistan imports are shooting up. How do we tackle this?'

Within that first week I had done extensive exploration in Pakistan and had a fairly good idea of what was hampering their dairy industry. I said to the Minister, 'Your Excellency, you have called me here to make your country self-sufficient in milk but I think you must be joking. How can you possibly be serious, when as the Finance Minister you did not impose any import duty on milk powder? Do you realise what it means when you don't put a duty on imports? It means that your Pakistani farmers and their cattle, with their unpredictable agro-climatic conditions, with their precarious fodder situation, must now compete with the New Zealand farmer and his magnificent Holstein cattle, with their agro-climatic conditions which are so marvellous that if it doesn't rain once in three days they call it a 'drought'. You should have imposed, at the very minimum, a 50 per cent import duty.'

To that the Minister replied: 'I had done exactly as you say and put a 50 per cent duty but I was forced to withdraw it within ten days. There was nothing I could do.'

'Forced, your Excellency?' I asked in some surprise. 'There are only six importers of milk powder into your country. Do you mean to say these six people forced you to sacrifice the interests of millions of your farmers?'

'Yes,' the Minister replied. 'That is exactly what I am telling you.'

'Then Sir,' I continued. 'I have only one more question. I have spent a lot of time in your market places this week. I see that there is a lot of milk powder from various parts of the world available in your shops. Let us assume that these six importers do not know that there is a world surplus; let us assume that they do not know that Kurien gets commodities free for his country; that they think your country is so rich that you can afford to buy it. Or, let us even assume that perhaps they don't know how to bargain – which is strange because if there is one thing all of us Asians know, it is how to bargain. Or, let us also assume that they are scoundrels who are building up their numbered accounts in Switzerland, by getting inflated invoices. Even then, Your Excellency, the price of milk powder in your country should be half of what it is. Who takes the other half?'

His Excellency threw up his hands heavenwards and exclaimed in disgust: 'This is the curse of my country!'

I knew at that moment that the Anand experiment in dairy development was not going to work in Pakistan – not until things changed drastically in that country and there were people in decision-making positions with the political will to bring about that change. My hunch was proved right.

Our team tried to put the systems and structures into place but the cooperative pattern simply did not take root in that country. Despite the disappointment that our attempts had not materialised, for me Pakistan was an extremely interesting experience and I met some wonderful people during my brief stay there. One of them was Brigadier Injaz Hussein, who was the Secretary Agriculture, Punjab.

One evening he took me to his house for dinner. When we

entered his house, he first locked the front door. Then he took me to his bedroom, went to his cupboard, opened it and pulled out a large album. He showed me scores of photographs – of him on horseback; photos of him receiving an award from the Queen of England and so on. Brigadier Hussein informed me that he was a very accomplished horseman. Then he confided, 'Since you now know that I'm a famous equestrian, what I want you to do, Dr Kurien, is to give me an opportunity to come to the Asian Games and be a judge for the equestrian events. You can see that I'm qualified to do this.'

I said to him, 'Brigadier, why are you telling me all this? I am just a simple dairyman. I am not running the Asian Games.'

However, he told me he had great faith in my ability to get this done and requested me to put in a word in the right quarters. So eager was he that I did not have the heart to disappoint him and on my return to India I did speak to some people in the right places. Brigadier Injaz Hussein arrived here for the Asian Games. He attended the games, presented some medals to the winners for tent pegging and was in seventh heaven. A very warm person like so many others I met in Pakistan.

My visit to Pakistan only served to make me more puzzled than ever before about the continuing and meaningless hostility between our two countries. For me, the only rationale for the Partition has always been that it was a plot of the advanced nations to make us feeble, and my visit to Pakistan further strengthened that belief. I was treated so well in Pakistan that when I returned I told some of my acquaintances in the government in Delhi that I wished they could take a lesson or two in this from the Pakistan government.

Some years later, another neighbouring country evinced an interest in seeking our help to improve their dairy industry. In the mid-1990s, Prime Minister P.V. Narasimha Rao asked me to lunch one day and informed me that the Sri Lankan government was keen to set up an Anand pattern milk cooperative there. Narasimha Rao wanted NDDB to assist them in this. Sri Lanka's President, Chandrika Kumaratunga, invited me to visit her country with my team in order to initiate the process of making Sri Lanka self-sufficient in milk.

In 1997, in collaboration with the Sri Lankan government, we set up the Kiriya Milk Industries in which NDDB had a 51 per cent stake. I was appointed its Chairman. Kiriya Milk Industries was to be the NDDB of Sri Lanka and we had hoped to make that country self-sufficient in milk in a decade.

When I first went to Sri Lanka, the government provided me with massive security. Not being used to having armed men follow me around, I was extremely uncomfortable with this. The first thing I said to the President when I met her was, 'Madam, what have you done? I have all these people in uniform following me around and I'm not used to this. I've always believed that when one does a job one takes whatever risks that go with it. After all, I am just doing a job like any other job – a low-risk job of building cooperatives. And incidentally Madam, I can speak Tamil so the Tamil Tigers will not kill me.'

President Kumaratunga said to me in all seriousness, 'Dr Kurien, I am not trying to save you from the Tamil Tigers. I am trying to protect you against other dangerous people. Do you know that here in my country, anyone can hire assassins for Rs 50,000?'

Once again, the ruthless attitude of foreign powers and their people was brought home to me. They get what they want by any method. Once more I was motivated to take them on. The problem with so many of our developing countries is, of course, the rampant corruption. The foreigners who forcibly occupied many countries and created multinational companies are convinced that 'the natives' are corrupt and can be bought. We see this widespread corruption in our own country too, but a point to be noted is that even in the midst of all this corruption, it is possible to create an Amul, it is possible to set up an NDDB. It is certainly not easy to do so but it is possible if you have at the helm very strong-headed and stout-hearted people who will not bow and will never compromise the larger interest of the common man. I had hoped to find such people in Sri Lanka, too, but it proved to be far more difficult than I anticipated.

At that time, the Sri Lankan representative of the New Zealand Dairy Board was a very influential man, one of the country's most renowned industrialists. I was reliably informed that he also happened to be extremely close to the President. This would prove to be a stumbling block because I knew that the New Zealand Dairy Board would certainly set off the first volley of opposition to Kiriya Milk Industries. When I asked the President about the industrialist she told me very honestly that he was, indeed, one of her friends.

'Madam, do you know what will happen?' I asked her. 'You call me here and you tell me, "Make my country self-sufficient in milk". What is the implication of such self-sufficiency? It is, first and foremost, that the New Zealand Dairy Board will have to wind up and go home. If I am to succeed here then there's no place for them.'

'Yes, I understand this,' replied the President very calmly.

But things were never easy in Sri Lanka. The vested interests were far too deeply entrenched into the system there. So much so that when the Labour Minister of Sri Lanka went to Geneva for a conference organised by the International Labour Organization, it was Nestle that sent a car to receive him and look after him while he was there. The Minister – Tondamman – a Tamilian in the Sri Lankan Cabinet – was an interesting man. He started his life as a tea plucker, then rose to become a labour leader of the tea pluckers and finally became the Labour Minister. I interacted with him quite often since Kiriya Milk Industries, predictably, ran into a fair amount of labour problems. We got to know each other fairly well and one day he told me quite frankly: 'Dr Kurien, I am a Nestle man. Don't ask me to do anything against them.'

Apart from this, unfortunately, the Minister for Trade – who was put in charge of the Kiriya Milk Industries – and I never saw eye to eye. It seemed to me that he stalled every move that we attempted to make. President Kumaratunga finally understood that I could not work with him and she very helpfully issued a notification transferring the Kiriya Milk Industries from the Minister for Trade to the Minister for Finance. However, even that did not work and the project to cooperatise Sri Lanka's dairy industry did not take off as many of us had hoped. Finally, in 2001, the President said that perhaps it was time that the Kiriya Milk Industries cut its losses and that we should instead think of bringing Amul into Sri Lanka in a big way.

It was evident that the authorities had to bow down to the extremely powerful multinational lobby. I cannot forget how upset Nestle was when Sri Lanka's President first invited us to replicate Anand there. Their representative in India came to Anand to see me with a shocking suggestion. 'Why don't you give me half of Sri Lanka and you take the other half?' he asked. 'You can choose which half you want.'

I was quite taken aback. I said to him, 'You are a great and famous multinational. What is a small Sri Lankan market to you? I am just a little cooperative. You compete with us – you destroy us if you can – but I cannot give you half. I want the freedom to go wherever I like.'

'How can that happen?' he said. 'You know that cannot work. And we hear that you are now planning on going to Kenya and Ethiopia too ... .'

'So then maybe it's time you planned to get out of these countries,' I retorted. 'Maybe you should start thinking of selling in Switzerland for a change!'

'All right,' he said angrily. 'We'll see who wins.'

I assured him that with the money power they had behind them it was a foregone conclusion that they would win, so they had no reason to feel insecure. As far as multinationals are concerned, money always was and always will be their only God. I once told the Chairman of Nestle during a meeting, 'I've been in this game for fifty years and I know your modus operandi well. Your problem is that in India you're running into people who know more about dairying than you will ever know. Your problem is that there is a Kurien here and you are unable to find out what his price is, so you're unable to buy him out, which is what you'd normally do. But you can't buy me out; you can't buy off Amul. Keep in mind that all your usual, unscrupulous procedures that bring you success everywhere else will not work here.'

Multinationals were not the only ones we antagonised. When Amul began exporting some of its products, some of the advanced dairying countries, too, were riled. Among them was New Zealand because in some of the countries Amul began jostling for market space against the New Zealand Dairy Board products.

Their supercilious attitude was epitomised by the High Commissioner of New Zealand who arrived in Anand one day, with the express purpose, it appeared to me, of abusing me. She sat across my desk in my room and said to me: 'It's all very fine that you are producing milk and selling milk. It's also acceptable that you are producing milk powder and selling it here. But now it has come to our knowledge that you have started exporting your products to countries where we sell ours. You cannot do that. That is our market!'

Although I have never stood for such arrogance, particularly from an alien to my land, I forced myself to remain calm, in deference to the fact that I was talking to a woman. 'Madam,' I responded very politely, 'quite frankly, I did not know that the world market is your private property and we should not encroach upon it.'

The lady was livid that I had dared to answer her and she grew ruder and more abusive. Still reigning in my temper, I suggested to her that it might be better if she left my room in case I said something I should not. But she persisted with her obnoxious remarks and I could not control my tongue any more. I said to her: 'Madam, you must understand that you come from this little country called New Zealand. If all of us Indians decide to get together and spit on your country, your country will get drowned in our spit.'

Whereupon, the lady got up and strode out of my office, if possible, even more incensed than when she had walked in to berate me. I was later informed by some friends in Delhi that she angrily narrated the incident over and over again in the capital's diplomatic circles, saying: 'That Kurien, he is a mad man. He threatened to spit on me!'

Life of Service

ALTHOUGH MY ONLY SALARIED JOB WAS AS GENERAL MANAGER OF THE Kaira Cooperative Union, I wore several hats in my lifetime. I often took on other responsibilities, whenever there was a problem somewhere and I was asked to step in. I was made the Vice-Chancellor of the Gujarat Agricultural University in 1984; in the 1960s, I was appointed Chairman of the Gujarat Electricity Board (GEB) where I had been a member for a few years.

During my tenure as Chairman, GEB, I realised that the structure of our electricity boards was highly ineffective. The Electricity Act had visualised the structure for a state board as one that would have some sixty thousand employees. I saw that such a mammoth bureaucracy could never be efficient for it is extremely difficult to remove corruption – or even prevent it – from entering such large institutions. Unfortunately, with our antiquated labour laws, it is difficult to promote someone for doing good work or punish someone else for bad work. It is impossible to deal with such huge numbers and these institutions are, therefore, destined to become inefficient and unproductive.

I recognised that all I would do sitting on the Chairman's seat at GEB would be to defend that scoundrel – my wireman – who was up to all kinds of mischief in some village. He would constantly indulge in petty crimes like giving illegal power for marriages, illegitimate connections, bypassing meters, stealing power, while doing this, at no time giving any payment to the electricity board. Such a system could never work. It would be almost impossible to ensure that an employee of the electricity board in a village is sensitive to the needs of the farmers and responsive to their demands. He would assume the role of the petty rajah of that place and behave as if he owned the village. In fact, I discovered to my horror, that my wiremen were teaching the farmers how to steal electricity.

However, there was not much I could do then, except prepare a detailed report of the shortcomings of the state electricity board, and note down some recommendations to revamp and improve this structure which had encouraged corruption and led to immense power losses. I suggested an alternate structure that I felt would work far better. It envisaged forming village-level power cooperatives at the initial stage and then forming an apex body at the district level – much as we had done with the milk cooperatives.

My term as Chairman of GEB lasted only for about eight months. As Chairman of GEB I needed to go to Baroda three days a week and the remaining days I spent in Anand, or travelling. I could not manage two full-time jobs and therefore decided to relinquish the chairmanship of GEB. Some two and a half decades after this, Gujarat's Power Minister came to meet me and said: 'Dr Kurien, everything you wrote on that file twenty-five years ago is true. The entire structure has become totally useless. I am not saying that GEB is the worst in the country but it is certainly not doing good work and I want to correct that. Why don't you think of bringing about the changes that you had suggested as Chairman of GEB? The Gujarat government will be extremely grateful if you would dismantle the board and rebuild it.'

'But what about the present Chairman of GEB?' I asked. 'What does he feel?'

The Minister said that he would ask the Chairman to meet me

and we should discuss the issue. The GEB Chairman and his officers did, indeed, have a long meeting with me and there was a consensus that a drastic change in structure was needed. They felt that for the good of the Gujarat state – and particularly for the benefit of the consumer – a new structure had to be evolved for the electricity board.

I gave them the blueprint for the new structure that I had envisaged as Chairman of GEB. I suggested that we should put one meter in a village and appoint a power committee that would be elected by the villagers. This village would then be charged according to the meter. The pattern of power distribution within the village would entirely be the villagers' decision. The elected committee would appoint a wireman and the people themselves would manage the power. But the board would charge the village by that one meter. If the village failed to pay its dues, power would be cut and no MLA, or MP, could force the board to reconnect power until the payment was made. According to a curious code of ethics we seem to have evolved in our country, stealing from the government may be honourable but stealing from your neighbour is dishonourable! Nobody would steal power because they would be watching each other and, automatically, power thefts would cease and power losses would stop.

When a number of such villages had formed elected committees, they would establish a union and a sub-station would be given to a union of such societies. It would be the responsibility of this union to appoint engineers needed to manage the sub-station. Thus, the board could go back to its job of generating power, leaving the distribution and management of that power to the people. It has always been my belief that once the tools of development are placed in the hands of the people, it becomes genuine development. As things stand, we do not have people staying on to look after the generation of power at our powerhouses because there is no money in it. All the money is made in the distribution of power. That is where corruption takes place. As a result of this, our powerhouses are neither properly maintained nor run. Once corruption in the distribution of power is stopped, things would automatically change.

My idea appealed to the Chairman and senior officers at the GEB. I emphasised to them that this experiment was important not just for Gujarat but for the entire country. I requested them to give me one district – Kaira district – with which to launch this experiment in power cooperatives. I warned them, however, that if they agreed to this, then GEB would have to withdraw from there, for Kaira district would build its own power station and the villages would take the responsibility to distribute the power. They heard me out and said that they could supply the power to the district but I said we would not take their power because we could not expect people to pay good money for rotten apples. For power to be useful to farmers, for it to serve as an instrument to improve the quality of life of our villagers – and particularly of our village women – it had to be of constant voltage, it had to be reliable and it should be reasonably priced. For all this to happen, we could certainly not rely on power supplied by the state electricity board. We preferred to generate our own power.

Moreover, I was quite certain that GEB employees would sabotage my experiment. Our own power generation was essential to preempt that eventuality too. We would put up a couple of combined systems of about 220 megawatts (MW) – which is a fair size – and the electricity board would be a standby. If Kaira district generated more power than we could consume, we would sell it to the GEB; if we needed more power then we would buy it from the GEB. Or, we could simply keep an account and exchange power when needed. This, in essence, was my idea for an experiment in setting up rural electricity cooperatives. At that point discussions with the National Rural Electric Cooperative Association of the US and donor organisations, for a grant of Rs 700 crore for a 220

MW plant and rehabilitation of power lines, were in an advanced stage. We felt it would be a path breaking project and if it worked, as it ought to, it could be replicated in all the states.

The GEB approved of my idea and asked me to send them a proposal. The Gujarat government, too, under the chief-ministership of Chimanbhai Patel, approved the proposal I tabled. The Chief Minister then sent my proposal to Delhi with a letter saying that the state wished to try out this system. I was called for a meeting in Delhi to discuss my proposal.

Predictably, the proposal met with intense opposition because the future of many bureaucracies called electricity boards would be adversely affected by it. I was persistent. I met the Chairman of the Central Electricity Authority and a number of other senior bureaucrats in Delhi but none of them viewed it positively. Finally, the Union Cabinet Secretary shot down the proposal. I felt rather dejected that I had failed to demonstrate what could be achieved by bringing power under the cooperative sector. I was certain that it would have dramatically decreased corruption, mismanagement and the unending power thefts that are routinely glossed over as 'transmission losses'.

Had I succeeded in convincing the powers that be, at least Kaira district would have seen efficient generation and distribution of power by consumer cooperatives. But once again, bureaucracy got in the way of development. This is the reason I have always had little patience with bureaucrats. Time and again they have revealed to me that they exist for their own power and pelf; they do not exist for the good of the country. In my rare moments of extreme despair I actually wonder why Kosygin's advice to me should not be heeded – that some changes can be brought about only by revolution and by beheading some people! This was certainly one struggle I gave up.

As far as I know, Chimanbhai Patel did not succeed in changing the minds of Delhi's bureaucrats either. However, a few years ago, two IAS officers – a Joint Secretary in the power ministry in Delhi and the other, a Joint Secretary in the power department in Gujarat – came to my office. They told me that they had reviewed the working of electricity boards and had come across my old proposal and felt that my ideas were absolutely correct. Since the power ministry was now agreeable to trying out my ideas, they requested me to put the proposal into action. I told them that it would take at least five years to get the experiment off the ground and I was not sure that I had five years left with me. Even if I did, I certainly did not have the energy to do so at this age. But I assured them that if they could find someone to translate my ideas into action, I would help them to the best of my ability. While I do feel heartened that my proposal is not buried yet, the apprehensions remain because the ball is still in the government's court.

In the Act of Parliament which created NDDB, the brief given to the dairy board extended beyond merely 'milk' and stated that the board would take on responsibility for 'milk or any other commodity that a state or central government may ask it to handle'. NDDB really started functioning like a development body for cooperatives and over the years the dairy board thus became involved in a variety of projects. People came to us looking for help and wherever exploitation of farmers took place, the dairy board tried to step in.

In my years as Chairman of NDDB, there was one experiment in cooperatisation which failed. One day in 1987, the Chairman of a milk cooperative in the Rann of Kutch visited me at Anand and narrated the heart-rending plight of the salt farmers of his region. About 60 per cent of India's salt is produced in the state of Gujarat and yet most of us here do not know of the harsh lives of the salt workers. Their situation, particularly in the Little Rann of Kutch, is very depressing. The salt workers have settled down in this arid, merciless desert where there is not a tree in sight. The villagers dig a hole in the ground, pump the water up and make a saltpan on some two hectares of land and farm salt. They work there for ten months a year producing salt. But they have to buy water from the merchant, to whom they ultimately sell the salt they produce. To get diesel for their pump they have to again depend on the same merchant. The salt worker finally gets two paise per kilo of salt from the merchant. There are no trees, no shelter for them, no schooling for their children. Long hours of working barefoot in the pans saturates their legs so much with salt that these farmers cannot even have a satisfactory cremation after death because their lower limbs do not burn. It is a miserable existence.

I confess that I was so depressed on listening to this tale of horror from the cooperative's Chairman, and so dispirited to see that, even after decades of independence such exploitation of our poor people continued, that I simply told the Chairman to leave because I did not want to hear anymore. I told him there was nothing we could do for the salt farmers – that we had enough problems of our own to tackle at NDDB. Two days later, very tragically, this Chairman died in a car accident. I was dismayed that I had been so unsympathetic during our last meeting and I thought that as a tribute to his memory, NDDB should enter the salt sector.

We began the task of organising the salt farmers of the Little Rann of Kutch. We started improving the quality of the salt and marketing it. During the first year we managed to produce and market 10,000 tons of salt. In the second year it went up to 25,000 tons, in the third year to 45,000 and then up to 100,000 tons. We tried to bring power through solar cells; we tried to provide medical facilities, water for drinking and bathing. It was a project that aimed to improve the overall quality of life of these salt farmers. However, within a few years we realised that we were able to market the salt through the cooperatives only to the extent so as to get the salt farmers about five paise per kilogram. This was marginally more than what they were getting earlier. By this time the Tatas began marketing iodised salt, which was of very good quality, and we felt that our project was not working out as it should. It was not making a basic difference to the lives of the salt workers. We had to, even if reluctantly, give up trying and wind up the project. Till this day, I look back at this failed experiment with an acute sense of helplessness.

~

When our country fought for its freedom from British rule, it evolved leadership of a quality that any nation anywhere in the world would have been proud of. We produced leaders of tremendous calibre, leaders with a nobility of purpose, not only from Gujarat – in the form of Gandhiji and Sardar Vallabhbhai Patel – but from every state in India. It is dismaying to see that we no longer seem to have that uplifting nobility of purpose to pursue, and we are producing increasingly inferior leaders. I have dealt with the government for over fifty years and I have found that grace in public life has become rare. Today in the frantic race to acquire power and money, vulgarity reigns supreme and the devil take the hindmost.

Yet, when we look around, we see that we are dependent almost entirely on our political and parliamentary system for future leaders. Since these structures now seem interested only in power and money, the quality of the leadership that emerges through them is, to say the least, ignoble. Therefore, today, perhaps more than at any other period in the history of our nation, there is a dire need to create a new generation of leadership through a plurality of truly democratic structures. What we at Anand and at NDDB tried to do was to create this plurality of democratic institutions at the village level, underpinning our democratic structure in Delhi or in the state capital. It is through such democratic structures, which permeate right down to the village level, that a dynamic future leadership of this country will evolve.

Take the case of Amul dairy. The elected Chairman of this dairy is a man who provides leadership to one of the largest food businesses in the country, and when he learns to handle such a substantial business, he becomes fit to hold any political position. Without having these schools and colleges through which our future leaders can emerge, our democracy itself cannot get strengthened. What is democracy after all? It is certainly not what it has, unfortunately, come to mean in our country: a government of the bureaucrats, by the bureaucrats and for the bureaucrats. This brand of democracy has no space for people. True democracy will emerge only when we allow the people to manage. And only when the people begin to take control of their lives will rural development gain momentum, when goods and services produced by rural areas will get better terms of trade than goods and services produced in the cities. Only when our farmers are involved in the processes of development will they be able to command their destiny. True development is the development of women and men. What we have to recognise is that 74 per cent of our population are rural people who depend on agriculture and if we want our country to develop, we have to ensure that these farmers and rural people develop.

As the Chairman of NDDB who always fought to keep politicians at bay from cooperatives, I was often questioned about how sugar cooperatives have worked despite being embroiled in politics, especially in Maharashtra. I have always looked at this positively. It is a fact that the sugar cooperative sector – which NDDB has never touched – is a success in Maharashtra. It is also a fact that the intensive dabbling of the sugar cooperative leaders in politics in Maharashtra has caused many a raised eyebrow.

However, the view I have always taken of this is that it is far better that sugar cooperative leaders control local politics rather than have powerful companies controlling it, because the cooperative leaders will certainly be more sensitive and responsive to the farmers' needs.

At NDDB we recognised that one of the biggest problems of Indian agriculture is that it is very difficult to fine-tune the production of an agricultural commodity exactly to match the demand. Production, being dependent on nature, will fluctuate. In such a scenario, it is easy for the traders and businessmen, whose only objective is to make money, to exploit the farmers. That is why the structure that NDDB promoted was a cooperative, non-exploitative structure, which eliminated the greedy people between those who produce the agricultural commodity and those who consume it. We believed that what worked through the Anand pattern with milk would also work for other agricultural commodities.

Therefore, we tried it in the edible oil sector and later, also with fruits and vegetables. It was Prime Minister Indira Gandhi who urged us to look into the latter area. During the period when Agriculture Minister Rao Birendra Singh was giving me a rough time, and I had sought a meeting with Indira Gandhi to talk about the difficulties I was having working with the Minister, she said to me, 'Dr Kurien, before we address your problem let me talk to you about my problem.' Then she proceeded to tell me about some fruits and vegetables she was growing in her own ten-acre plot. 'I get only one rupee a kilo for the vegetables I grow, but when I buy the same vegetable in the market I have to pay six rupees per kilo,' she explained. 'Dr Kurien, who takes those five rupees and why?'

I explained to her the problems of Indian agriculture and the solution to eliminate the middlemen. In a way, out of this inquiry was born the fruits and vegetables project of NDDB, which became operational first in Delhi in 1988 and thereafter, was replicated in several other cities. Like all the other NDDB projects, this, too, provided a direct link between fruit and vegetable growers and consumers.

The dairy board extended the same principles of cooperation to the realm of forests, too. For years, we had been hearing that about five million hectares of our forests are denuded each year. Political leaders discussed at length methods to reforest this area. But their methods revolved largely around creating a forest department, a forest minister and secretary and the unwieldy bureaucratic system. No doubt a large number of people would have been employed but it is debatable whether trees would get planted and forests would grow! On the other hand, NDDB believed that if we could leave it to the people to plant trees, and if we could arrange it so that the benefit of the plantations would go to those who planted them, then perhaps, our barren lands could be reforested.

In 1986, we initiated a pilot project involving tree growers' cooperatives. The aim, once again, was to create a people's institution – this one, to meet the fuel and fodder needs of rural people by growing trees and grass on common and degraded wastelands, as well as on marginally productive private lands around the villages. As we had anticipated, this pilot project was a huge success and, from that, rapidly evolved the National Tree Growers' Cooperatives.

While NDDB has done much for cooperatives across the country, the task is by no means complete. I believe that farmers should be permitted to handle all the agricultural commodities they produce rather than have the government trying to run the show. I do not think, for instance, that we need a Cotton Corporation of India to handle cotton, or a Jute Corporation of India to handle jute and frankly, I do not think we need a Food Corporation of India to handle food. If, instead, the farmers are allowed and encouraged to handle their produce themselves, not only will there be a better handling of these commodities but farmers will become better and more responsible citizens of this country.

Many of the obstacles that we faced at NDDB arose because the laws ruling cooperatives in our country tied our hands. It is a matter of shame that even today, in India, 'cooperative' is not a good word. Our cooperative law is still archaic. The bureaucrats and politicians dominate it; it has no honoured place for professional managers to function in. Once such impediments are removed, once we can inject professional management into cooperatives, and we can depoliticise them to the extent possible, only then will cooperatives flourish the way they ought to.

Cooperatives that are not government controlled will certainly be better equipped to take care of competition. Provided that we are allowed to compete on a level playing field, we will manage to hold our own. After all, despite the entry of multinationals in our country, who is the biggest manufacturer of baby food today? It is Amul, a cooperative.

While I think that privatising and removal of bureaucratic control from the economy is desirable and I welcome that, I object to the sequencing. I would have been so much happier if our former Finance Minister, Manmohan Singh, had first liberated agriculture before he liberated industry. If we leave agriculture in shackles and liberate industry, industry will only further prey upon it. I am one of those who firmly believe that our cities thrive at the expense of our villages; that our industries exploit agriculture. Unfortunately, the government seems to have forgotten that there is this important sector of the economy called agriculture, on which the majority of our population still lives. The politicians and bureaucrats forget that there is, in fact, a silent majority – the millions of farmers who have no lobby, no contact men to speak for them in Delhi, and therefore, their voice never reaches the government. But that does not mean that they do not exist and neither does it mean that they are not important. If government does not pay heed to this, there will be a terrible imbalance and, in the end, a heavy price to pay.

The bottom line is that the government must govern, in every sector. The government need not nationalise banks but government must see that banks do not defraud people. The government need not run dairies but the government must ensure that the private sector puts out good quality milk at reasonable prices. That is governing. So, let the government get out from places where it should never have been in the first place.

A LOOK BACK

I CAME TO ANAND A LONG TIME AGO, NOT BECAUSE I WANTED TO COME HERE but because I was compelled to. I was under a contract with the Government of India, which had paid for my education in the US. I was an engineer who was primarily interested in physics and metallurgy but studied dairy engineering under duress. Then I was forced to go to Anand to join the government's research creamery, and when I later joined the cooperatives of dairy farmers, I was not really trained in this field.

I was city-bred with no affiliations with rural India. I had a commendable academic record but no particular knowledge of farmers or agriculture. A number of events followed and today I can look back and say that because of this string of accidental events, the dairy industry in India is not what it used to be. Had these 'accidental events' not taken place, perhaps our dairy industry would not have developed to the extent it has. And I am certain that I would not have been the person I am.

The realisation that we cannot depend upon an 'accidental' Kurien, but we need to actively create better Kuriens by carefully selecting and training them, got me thinking. As the implementation of Operation Flood proceeded I could clearly see that most graduates from the Indian Institutes of Management (IIMs) were not quite suited to the kind of work that needed to be done at the cooperatives. Our IIMs were training students for the corporate and not the cooperative world. We needed an institute to train youngsters specifically in rural management. We needed an institute to create and strengthen commitment to serving our rural people with dedication and honesty.

My reasoning was even if only a handful of students each year from such an institute could grow, evolve, stride out into the world and, perhaps, build some more Amuls, what a boon it would be for our rural people and for the nation. This vision, to fill a growing need for professionally trained rural managers for the cooperatives and the NGO sectors of India, was the beginning of yet another dream – a dream which ultimately took the shape of the Institute of Rural Management Anand (IRMA) in 1979.

Working with me closely to create a blueprint for this institute, was the inimitable Ravi Matthai, the then Director of the Indian Institute of Management (IIM), Ahmedabad, and an ardent supporter of our work at Anand. He also happened to be my cousin – he was John Matthai's son.

Vikram Sarabhai, who was the first Honorary Director of IIM Ahmedabad, once met Ravi Matthai at IIM Calcutta. He was quite impressed with Ravi and persuaded him to come to IIM Ahmedabad as its full-time Director. On moving to Ahmedabad, Ravi would come to Anand to stay with us on weekends. He had a brilliant mind and was a delightful fellow and Molly and I were very fond of him. I used to argue with Ravi and criticise the IIMs saying, 'All your students have their education highly subsidised by the government and yet the minute they graduate they push off to the US and the UK. Don't they feel any obligation to serve their country, at least for a couple of years? Obviously, you are not motivating them in the right way.'

Ravi suggested that I join their board and deliver these 'lectures' there. So I joined the Board of Management of IIM Ahmedabad. At the first meeting, I delivered my lecture – about how we should motivate the students to feel that it is a matter of glory to work for our country; about how they should serve the nation for at least three years before they go in search of their own progress abroad. The other members of the board sitting around the table were some of Ahmedabad's top industrialists. One of them took his cigar out of his mouth, looked at me and said superciliously, 'So Dr. Kurien, you want our graduates to go and milk cows.' I stood up, returned his look and said, 'No, you continue to teach them how to suck on cigars.' And then, adding that I was resigning from the board, I walked off.

The next time Ravi met me, he said, 'Looks like my decision to take you on my board did not turn out right.'

I said, 'Yes, it did not turn out right. So what do we do now?'

He remarked that I would never be happy until I built my own institution. But I did not know anything about management education. I told Ravi, 'I can build the body, but who will provide the soul? Will you provide the soul to the institution? Will you join the board?' He said he would and thus IRMA was built with Ravi on its board. He was intensely involved in the entire process – in shaping the concept, the structure and the curriculum and in choosing the faculty. He admitted to me all the mistakes he had committed in building IIM Ahmedabad. It must have been painful for him, like criticising one's own child. But he was determined not to allow us to replicate those mistakes.

Unfortunately, Ravi passed away before the work could complete. When the IRMA library was ready, I invited Rajiv Gandhi to inaugurate and dedicate it to the memory of Ravi Matthai. We called it the Ravi Matthai Library.

When we designed Operation Flood, we made a provision of Rs 50 lakh for the training of human resources to manage all the cooperatives that were going to be established. The idea evolved into IRMA, ultimately costing us Rs 10 crore. But it was money fruitfully spent. We did not cut corners anywhere. IRMA has one of the most beautiful campuses in India and some of the best facilities for the students. Our expectations were extremely high and we knew that the standards of any educational institution largely depended on the director and the faculty of that institute. It was not easy to find faculty oriented towards rural India. Since we did not have a ready-made faculty with the type of orientation we needed, we had to initially borrow them from other educational institutes that were mostly oriented towards the industry. They needed time to adapt themselves to the needs of the cooperatives but we were patient. And our patience is paying off.

Today, many of our early students are in senior positions with cooperatives or NGOs across the country. No doubt some join the corporate world but I believe that even those who do so are better managers for, during their years at IRMA, they get a rare and valuable insight into the rural and agricultural reality of our country. I am sure that this experiment will, indeed, in the long run, bring about the transformation of rural India that we dreamt of.

Building a nation's dairy industry is a tall order. It was only through the commitment and perseverance of many people, including millions of our rural milk producers, that we were successful in using the cooperative strategy for dairy development. Insofar as my professional involvement and modest contributions to the development of the dairy industry are concerned, it all started in December 1950 when the Founder-Chairman of Amul, Tribhuvandas Patel, noticed my skills and persuaded me to stay on in Anand. Tribhuvandas instinctively recognised the value of professionals and from the very first day, placed his trust in me without reservation.

Tribhuvandas was an extraordinary man. He had been born into an extremely wealthy family in Anand, but had dedicated his life to the service of the nation, first as a freedom fighter and then as the untiring leader of Kaira district's dairy farmers.

Tribhuvandas's unfaltering support to me – in all the projects I initiated and through all my confrontations with the government and politicians – lasted during his entire lifetime. He was the one who ensured that my work as the Chief Executive Officer of Amul was always insulated from institutional and political pressures. He brought out the best in me with his inspiration and direction and convinced me that there was a special satisfaction in working as an employee of India's backbone, the farmers. His judgement of farmers always remained better than mine and whenever I disagreed with him on crucial issues, I was ultimately proved wrong. He was a real leader and a true builder who built with commitment, purpose and unimpeachable integrity.

Many people found it strange that, for so many years, a Keralite Christian had been regularly and unanimously elected Chairman of the GCMMF. There was a time when I was also Chairman of the Gujarat State Electricity Board. Another time I was Vice-Chancellor of the Gujarat Agriculture University. I doubt if in my own home state of Kerala a Gujarati would ever have been given these important positions of power. Yet I am aware that none of this would have been possible without Tribhuvandas Patel. He selected me as his senior executive, knowing that I was from Kerala and a Christian. Religion, caste, community did not matter to him. He only asked if I was committed to my work, if I was an honest man.

Unfortunately for all of us, even Gujarat does not produce any more Tribhuvandases these days. The Gujarati is an international figure. Gujaratis do business not just in India but throughout the world, which is all the more reason why Gujarat should not become clannish and narrow-minded. It is not in Gujarat's interest to become sectarian. The Gujarati should remain like Tribhuvandas – truly national.

In 1994, Tribhuvandas fell ill. Age, and a lifetime of dedicated work and sacrifice, with never a thought for his own comfort, had finally caught up with him. Towards the last stages he was in and out of the hospital – a place that he hated. Tribhuvandas knew that the end was near and he asked to be allowed to remain in his own house rather than in a hospital. Till the very end, from his own bed at home, he kept himself updated about the work at the cooperative, which had been his lifetime's commitment.

It was my misfortune that during the last days of his life I was called away for an important meeting in Delhi. When I called from Delhi to inquire about his health I was told that he was getting better. Therefore, I decided to spend one more day in Delhi – a decision I have regretted all my life. By the time I returned to Anand, he had already passed away. I never expected it to happen like this. My colleague, A. A. Chothani, who had always been very close to him, informed me later that Tribhuvandas had waited for me, quite certain that he was not going to live beyond six-thirty p.m. that day. This is exactly how it happened. A few minutes before six-thirty, he called Chothani near him and said: 'Tell Kurien that I waited till six-thirty. He has not come but I can wait no longer. Now you must tell him that I am leaving everything – the cause of the farmers and the cooperatives – in his hands.' Those were his last words. In Tribhuvandas's passing away, I lost my guru.

After Tribhuvandas died, the popular thinking was that he should be cremated at the Amul dairy grounds. I intervened, saying that I would not allow this. Tribhuvandas would have been the first to say that the Amul dairy did not belong to him, it belonged to the farmers. Besides, I explained, if we cremated him there we would have to build a memorial to him. This would set a precedent. With this decision I know I upset quite a few people but Tribhuvandas would have agreed with me. Finally, my decision prevailed and we took his body to the burning ghat where the last rites were performed.

~

I was just twenty-eight years old when Tribhuvandas made me the General Manager of Amul. It is an age when one believes nothing is impossible and one is ready to take on challenges. I believed in entrusting professionals with responsibilities at an early age, encouraging them to take initiatives, and correcting them when things went wrong, instead of penalising or condemning them.

While integrity and loyalty are core values, there are other values, too, which are a prerequisite to achieve success in any field. For example, the leader has to set a personal example and make others understand in what ways 'change' is going to be useful. I believe that professionals working in our organisations must have a clarity of thought combined with a passionate pursuit of mastery of their subject. I have always emphasised that large endeavours are only the sum of many small parts and, therefore, we must keep in mind not just where we are going but how we are going to reach there successfully. I also believe that a person who does not have respect for time, and does not have a sense of timing, can achieve little. To help employees foster these values in NDDB and the other organisations with which I have been associated, I have always given them challenging jobs, enough freedom to perform those jobs and adequate opportunities to prove themselves.

The involvement and support of the right people has always been one of the keys to the success of any mission. Mine was no different. Very early in my association with Amul, I was able to persuade H. M. Dalaya, one of the most knowledgeable persons about dairying in India, to join me. Much of the success of Amul and of the institutions inspired by it is in large part due to the efforts, the creativity and the wisdom of Dalaya. I learnt much from him that has stood me in good stead over the years.

I was also immensely fortunate that the home of one of our outstanding bureaucrats, H. M. Patel, was in Kaira district. He was a giant, not only in our civil service and national political life, but, even more so, in his commitment to building the institutions responsible for real development. It was through his efforts that Kaira is endowed with educational institutions, health facilities and many other parts of an infrastructure now taken for granted. From H.M. Patel I learnt the art of administration. More importantly, his example of unflinching integrity and commitment to principle has remained with me as a living inspiration.

The NDDB, which it was my good fortune to chair for thirty-three years since it was founded in 1965, vastly benefited from the services of hundreds of truly outstanding young women and men. They worked with courage, with diligence, with unparalleled commitment, with little financial reward and less fame. I learnt a great deal directly from the experience of some and indirectly from the experience of many.

Amrita Patel, who is the present Chairman of NDDB, was a colleague and a close friend for many years. She joined us as a young veterinarian and I was soon impressed with her organisational and managerial skills, which she displayed while carrying out her responsibilities during the important International Dairy Congress, which for the first time was held in a developing country, in New Delhi in 1974. I had, of course, known her earlier as the youngest daughter of H.M. Patel, but her abilities within the organisation led me to identify her as the person to succeed me as Chairman, NDDB. Patel had all the requisite professional qualifications for the job and I trained her for twenty-five years to take on the reins of a hugely challenging task.

There are also other innumerable friends and colleagues who, as my thinking evolved on a number of subjects, helped me greatly by their research and sharp minds. As I never tire of saying, my colleagues did all the work and I collected all the awards.

Finally, I owe much to my wife Molly, my guiding star, who has not just helped me to stay on course but has always been very understanding and supportive. She has always been there as a constant source of strength.

~

When I became Chairman of the National Dairy Development Board in 1965, there were no takers for the post. When I offered my resignation and stepped down in 1998, the bureaucracy suddenly came up with a host of potential candidates. Clearly, they felt that the time had come to see if they could capture this organisation with its corpus of Rs 3,000 crore.

The problem with most cooperatives in India is not that politicians become chairmen. It is that when they become chairmen, they use the cooperatives to further their own political ends. The beauty of all the cooperatives at Anand is that we have had chairmen of high integrity. Tribhuvandas Patel set the first example and we maintained that tradition very strictly.

What is the primary job of an efficient manager? In my book, it is to bring in and groom the right people on the team. Once this is done, the manager must then groom the successor most appropriate for the institution.

For three decades I was the Chairman of the NDDB and was reappointed every three years. Many political parties came into power and yet whoever came to power in Delhi reappointed me. Yet in 1998, when after careful consideration I decided to pass on my responsibilities and submitted my resignation, all hell broke loose. The Cabinet Secretary in New Delhi and the Principal Secretary to Prime Minister Atal Behari Vajpayee immediately appointed a search committee to look for a successor. Certainly, the corpus must have been a tempting factor, but that money belonged to the farmers. How could I have allowed it to go into the hands of anybody other than whom I had groomed, and in whom the farmers had complete faith? There I was, grooming Dr Amrita Patel as a successor for over two decades and they appointed a search committee!

They said they did not want Amrita Patel. They said they did not want somebody who had been groomed by me. But things were not that easy for the bureaucrats, largely because when I had been asked to draft the Act of Parliament, which set up NDDB, I had been rather clever about how it was worded. One of the clauses of the Act stipulated: '... the Chairman shall be professionally qualified in one or more specialities namely: agriculture, dairying, animal husbandry, rural economics, rural development, business administration or banking ... .'

That is where they ran into trouble. They knew that they would have to find someone professionally qualified. Yet, the person had to be an IAS officer. It is not at all easy to find an IAS officer who is thus professionally qualified, but finally they identified an officer from Rajasthan.

The minute I heard this I went to see the Prime Minister, to whom I always had access only because I represent ten million farmers. Political leaders may not do what I ask them to do, but I was confident that at least they would not take a decision without having heard me out. When I met the Prime Minister as well as the Home Minister, L.K. Advani, who was responsible for appointments to senior positions, I explained to them that giving the NDDB's reins in the hands of the bureaucracy would be a recipe for its ruin.

As I was walking in to meet the Prime Minister, I met a Cabinet Minister. He called out to me, 'Hey Kurien, what is this I hear – that you're being replaced by an IAS officer from Rajasthan?'

I told him that was true and I was going to try and persuade the Prime Minister not to do this. The Minister said: 'I know this person. He is definitely not the right kind of person for NDDB. He will loot the place.'

I requested the Minister to tell the Prime Minister this too. I do not know if he did. But there were several other senior and respected politicians who tried to dissuade the government from appointing an outsider. Among them was former Minister of Agriculture, C. Subramaniam, who wrote a strong letter to the Prime Minister, saying that if a politician or a bureaucrat was brought in as Chairman of NDDB, it would be disastrous for the organisation. Instead, the government should allow the person whom I had groomed to take over as my successor. It helped that Amrita Patel had the firm support of the Home Minister, who had great respect for her father, H.M. Patel. Once more, while many bureaucrats tried to put a spoke in the wheel, there were others who supported us. Finally, Amrita Patel was appointed as Chairman, NDDB.

~

There is a saying in Malayalam, my mother tongue, that the jasmine blooming in one's own backyard never smells as sweet as the jasmine in somebody else's garden. How often this is true. I would be dishonest if I denied that it caused me some anguish when recognition from my own country for the work I had done came only after I received international acclaim.

Yet, the awards I treasure most are the ones my country has bestowed on me and those that I have received from the Indian farmers. There is always a great satisfaction in receiving awards but I have never been able to take personal pride in them. I have been acutely conscious that these awards, in fact, do not honour Verghese Kurien as much as they honour millions of dairy farmers of our country, who against all odds, raised India to the first rank among the world's milk-producing nations. The awards are an acknowledgement too, of the dedication of the tens of thousands of elected leaders of India's dairy cooperatives and the professional employees who have committed their life to serving our farmers.

The awards are a recognition of the contributions of hundreds of my colleagues at NDDB, Amul, GCMMF, IRMA and all our sister organisations, who vigorously and continuously supported my efforts with their exceptional skills, integrity and commitment.

The work we did at Anand to empower the poor inevitably faced opposition, some of it fierce. To some extent, recognition from other countries and the international awards I received gave a fillip to my determination and strength to continue that work. The first international recognition of this work came, in 1963, in the form of the Ramon Magsaysay Award for Community Leadership which was rightly shared with Tribhuvandas Patel and Dara Khurody. It was after this that my government deemed me deserving enough to receive the Padma Shri in 1965 and immediately following that, the Padma Bhushan in 1966. In 1999, I received the Padma Vibhushan, the nation's second highest civilian award.

These three national awards occupy the pride of place in my home today. All my other awards have been given for display at the Amul Museum which was recently set up in the house that Molly and I once occupied. More than one hundred and fifty awards and citations from across the world adorn the walls of that museum as testimonials to the achievements of India's dairy farmers.

While the recognition from farmers and the three national awards will always hold a special place in my heart, there are some others which, for one reason or another, have touched me. One of these was the Samanwaya Puraskar presented to me by Swami Sathwamithranand in Haridwar in 1990. I have never been sure why the Swami felt he should honour me but I did go up to his ashram in the Himalayas to receive the award. It was quite an exhilarating experience. An elaborate function was organised to present the award to me. On the dais, to my right, sat the Swami and on his right sat L.K.Advani. On my left was the leader of the Vishwa Hindu Parishad. In front of me was a spectacular ocean of saffron – hundreds of sadhus who had gathered there for the event. L.K. Advani spoke, praising me for my work, and I was then presented the award along with a cash prize of Rs 5,000.

Although I have never been a religious person, I attended this function because I thought it was significant that a Hindu religious head should single out a Christian to honour him for his work. In my acceptance speech I acknowledged as much; I said I felt privileged that the sages of the Himalayas should felicitate someone from the southern-most tip of India. I was a Christian by birth and yet I was being honoured by them and what this symbolised to me was extremely important – that whichever corner of the country I came from, whatever my religion, I was first and foremost an Indian. That is why this award was a particularly prestigious one for me.

I was deeply touched, too, by the Dr Paulos Mar Gregorios Award which I received in 1999. It said a lot for the liberal attitude and the tolerance of the Church which gave me this award, despite my public statements that I was an atheist and that the only time I went to church willingly was when I got married!

I was presented the Wateler Peace Prize Award of the Carnegie Foundation for the year 1986 and three years after that I received the World Food Prize for 1989.

~

I was Chairman and General Manager of GCMMF, where the rules demanded that the Chief Executive Officer retire at fifty-eight but could be given an extension for another two years. I had to follow the rules that I myself had laid out; after an extension, when I turned sixty in 1981, I retired as the General Manager. Although I continued as Honorary Chairman, my salary had to stop. People invariably asked me how I managed to survive without a salary and I replied only half jokingly: 'I win awards. Now only the Nobel Prize eludes me.' Indeed, I have won many awards, a number of which were accompanied by cash prizes, and those did manage to keep the home fires burning.

My attitude towards money has always been a very realistic and a utilitarian one. It springs from one of my deepest convictions, which I also tried incessantly to instil in my daughter Nirmala during her formative years, and later in my grandson, Siddharth. I remember often explaining to Nirmala that it is terrible to have too little money because you will not even have enough to eat and appease your hunger. But it is far, far worse to have too much money because then you will surely get corrupt. Our family, I think, was truly very blessed because we always had only just enough.

Even if Amul is the largest food business in the country today, cooperatives have never been able to match the salaries of multinationals because we have been conscious that it is the farmer's money that is being spent. The maximum salary I ever drew was Rs 5,000 as the Chairman and General Manager of GCMMF. I never felt underpaid and we managed with the money we got. In hindsight, I realise how difficult it must have been for Molly to manage household finances with that salary. Yet, she gave us a beautiful yet simple home, she provided us with two full meals a day and graciously looked after the unending flow of house guests in the early years of our marriage. We finally moved into our own house in 1989. It is a modest but lovely house in Anand, built by drawing fully from the provident fund that I had saved over the years.

One day, an acquaintance of mine – a senior officer in a corporate firm – came to meet me to complain about his unhappy fate. He seemed to be extremely dissatisfied with his work and his life in general. Listening to his laments, I thought that perhaps he felt he was not earning as much as he deserved and so I asked him the kind of a salary he drew. He told me he earned Rs 60,000 a month. I was amazed at this. I asked him why, with such a handsome salary, should he be so unhappy? He explained: 'It's not the money, Dr Kurien. The money is fine but nobody gives me any respect in my office and that bothers me terribly.'

I was immensely amused at the irony of the situation. I said to him: 'Look at me – I get hardly any money but I get a lot of respect. Now, my friend, which is a better situation?'

I may be old-fashioned in my thinking but I have always believed that it is only when you get less than you are worth, that you can look for respect; if you are paid much more than you are worth you will get no respect. The one thing that I have never had a complaint about is the amount of love and respect which my dairy farmers showered upon me.

There were times when this respect took on absurd but amusing dimensions. I remember, when I retired from NDDB in 1998, Amrita Patel very generously insisted that I should keep the NDDB Chairman's car to use as the IRMA Chairman's car. I refused, and instead, advised her that she should keep the car as the new NDDB Chairman, because of the aura of respect the car seemed to radiate over the years. There is a story behind this advice.

There was a time when Molly and I used to have a pair of magnificent Irish Red Setters to whom I had grown very attached. We had bought a new car for Nirmala and I wanted to take it for a test drive. I had already changed into my kurta-pajama for the night but as soon as I took out the car and opened the door, the dogs jumped in beside me. I drove towards the NDDB gate and one of the watchmen on duty stopped the car, not recognising me in my nightwear. The other watchman, however, peering in from the other side, where the Irish Setters were hanging their heads out of the window, enjoying the breeze, shouted across to his colleague: 'Nahi, nahi. Saab ka kutta hai, saab ka kutta hai! Usko jaane do.' Evidently, the saab was immaterial, but saab ka kutta was allowed in. Respect rubs off onto all kinds of things and creatures around you through some peculiar process of osmosis. As I said to Amrita Patel while refusing to take the car, 'At this stage, people will look at the car and say, "See, there goes the Chairman of NDDB!"'

~

I first learnt to play chess when I was six years old. Those days it was not easy to get a proper chessboard in my home town but my cousins and I were keen to play this game so that we turned creative and innovative. We would get hold of the hard core of a banana tree, which we would then proceed to cut and chisel into the shapes of the chess pieces. A piece of chequered paper served as the chessboard. We played the game avidly and constantly and as a result I became a good player. It was only much later in life that I realised the tremendous benefits of this classic game of strategy. Chess had taught me the skill of how to always remain a few moves ahead of my opponents.

Perhaps that is partly the reason that all through my years at Anand, I worked well in crisis situations. I always felt that one of the best things about being confronted by problems was that you could find solutions to those problems. In many ways, I think I was in my element when I was being attacked. In this, I fully endorse the view of my friend Vikram Sarabhai who often said to me: 'When you stand above the crowd, you must be ready to have stones thrown at you.' I accepted every controversy that came my way as a new challenge. It was because of this perseverance in the face of adversities, that my colleagues and I could make NDDB into the kind of organisation it became.

Many years ago, Prime Minister Rajiv Gandhi sent Sam Pitroda, a telecommunications expert who served as his adviser, to meet us to discuss his five technology missions. Pitroda in his address, that startled some of my colleagues, said, 'In order to get things done you have to be a bit mad.' And then pointing towards me, continued, 'You have an example in front of you of a man who came with a dream – a bit mad – and today we are all reaping the benefits of his vision, his ideas. We don't have to agree with him on everything or admire every little thing he does but the fact remains that what he has created is a dream that we all like to share. We need more and more people like him to create more dreams for our people.'

Perhaps he was right. Perhaps I was a bit mad to have ever believed that the dream I had to make our country self-sufficient in milk through our dairy farmers' cooperatives would actually be realised one day.

I have, by and large, followed my instincts in most decisions I have made. But I also recognise that I was successful in the endeavours I undertook largely because of the Government of Gujarat's enlightened policies in regard to dairy development. Having worked with various other state governments I always found that Gujarat was fortunate in the type of government it has always had. Regardless of the political party in power, the government of this state always allowed people's organisations to grow.

At this point some figures are, perhaps, apposite. Particularly for all those who doubted our motives at NDDB and for all those who felt that our farmers were not competent to achieve the unachievable. Milk production in India has grown at a healthy clip of 4.8 per cent during the last twenty years and India is now the largest milk producer in the world. Today, milk availability per capita is 214 gm per day, exactly double the figure of 1970 when we launched Operation Flood. About 66 per cent of rural households and 90 per cent of urban households consume milk. On an average 50 per cent of milk production is retained in our rural areas for consumption. The per capita household consumption of milk has increased by 23 per cent and 15 per cent in rural and urban areas respectively between 1987-88 and 1993-94. The cost of milk production in India is nearly 40 per cent lower than the corresponding figures for the European Union and USA, and compares very favourably with those of Australia and New Zealand.

I do not think, therefore, that it would be unfair to say that India's dairy industry has, indeed, come a long way. But our mission is still far from over. Dairying in India should now rapidly progress in quality and in terms of retaining its relevance as the country's largest rural employment programme. This is both a challenge and a threat – given that in this new era of an integrating world economy, the rules for international trade in milk and milk products are written by developed nations who believe in 'mass production' and not 'production by the masses'.

It is now over a decade since India embarked upon its journey of liberalisation and globalisation. In the days that have followed, most of our cooperatives have learnt to cope with three distinct disadvantages that a liberalising economy has conferred on them. Firstly, our cooperatives have learnt to develop without the benefits earlier available to them in a mixed economy. Secondly, our cooperatives have learnt to perform in the market place even though the shift to a free market economy has mainly benefited large capital, whether Indian or foreign. Thirdly, even though the labyrinth of rules and controls that regulate cooperatives remain generally unchanged, our cooperatives are finding innovative strategies to achieve their goals.

Essentially, our dairy cooperatives are becoming proactive and learning new ways of adapting to change, even as they eagerly wait for that logical stage of liberalisation when they will have a level playing field with their competitors. I can only say that when our cooperatives are able to perform and deliver in an environment that is still quite adverse, imagine what will happen when they are permitted to carry out their business with the same freedom and flexibility that is already being enjoyed by their competitors.

The dairy cooperatives have managed all this because they have done three crucial things. They have acquired and equipped themselves with the latest and most modern technology for milk processing and product manufacture. Then they have defined the standards necessary to achieve and maintain world-class quality. And, most importantly, they have put in place systems which will ensure that they consistently achieve the standards they have set for themselves.

After having worked for the cause of our nation's farmers for over fifty years, my vision for the twenty-first century remains the same. I sincerely believe that India can achieve the status of a developed nation and occupy its rightful place in the affairs of the world – hopefully within the first few decades of the twenty-first century.

But I can never stress enough that each one of us has a responsibility, as a member of our nation's privileged elite, to help bring about that stage. That responsibility is both to criticise and to correct. We must take responsibility for our nation's future; we must hold ourselves accountable for that future. It means that we must act not only as advantaged individuals but as concerned members of our society. It means that in all that we do, we must be aware of its effect for the greater good.

I have often claimed that I have had but one good idea in my life: that true development is the development of women and men. This idea took such a hold of me that I remained in this small, sleepy town of Anand for over fifty years as an employee of farmers. I was never able to give this up for what many call 'a better life'. These years have, without an iota of doubt, been the most rewarding years of my life. Over the years I have spoken ceaselessly of this idea, hoping to enthuse young women and men to adopt my passion as theirs. I have been fortunate that there have been many who took up the challenge. Yet I cannot help but feel some disappointment that, in the larger scheme of things, our policy makers and implementers (with honourable exceptions, of course) still believe that our nation's women and men are means, not ends.

Yet, if there is one conviction I hold as firmly today as I did fifty years ago and if there is one reason why I must remain optimistic about our nation's future, it is because I have seen over and over again that when the tools of development are placed in the hands of our rural people, and when their energy and wisdom is linked with the skill of committed professionals, there is nothing they cannot achieve.

Our greatest national resource is our people and too often have we neglected this resource. We have, whether intentionally or not, created the illusion that all resources are in the hands of the government; that is, the government and its agents who dole out resources. India would have been a far stronger nation had our people been asked to create resources. Our schools, our health services and many other community programmes would have been far better had communities raised the resources, created the facilities, employed the staff and held them accountable. But sadly, our communities have no say in these.

We have glorious examples in our country of what our people can achieve – and have achieved – by working together. There are cooperatives, there are citizens' groups, there are communities who have come together to volunteer their time and efforts for the benefit of others. All these initiatives have worked far better than the efforts of the government. And it is for one reason: that those who raised the resources, created the institutions and gave their energies to these endeavours, truly cared.

Of course there is a role for the government. But it is a more modest and focused role than what we see today. There is also a role for individual and group efforts in building a better society. All it takes is for people who care to come together, to inspire others, to involve others. But the first step must always be the recognition and conviction that the change can only begin with ourselves.

I look around my country and find problems that often seem overwhelming. Despite fifty years of freedom we have yet to solve the very real problem of poverty, whether rural or urban. We can feed ourselves – which is a great achievement. However, we cannot fool ourselves. Instead, we need to see that it is a precarious balance – one that is achieved only because, sadly, there are all too many of our fellow countrywomen and men who cannot afford enough to eat. Were they to enjoy the same diet as most of our privileged class does, it is questionable whether our national production would suffice.

While we can take some pride in the fact that our fellow Indians can meet any challenge and achieve success in highly competitive fields abroad, we must also feel a great sadness that all of that talent – talent that was nurtured at the cost of our own country, since all our academic institutions of excellence are heavily subsidised – is being placed at the disposal of others and not our own country, which so desperately needs it.

Our country faces great challenges. There is no doubt that today, our body politic is a cause of grave concern. There is no doubt that hundreds of millions of our people do not enjoy an acceptable quality of life and nor are they able to reach their potential to contribute to our nation. And yet, there is no doubt in my mind that as a people we stand second to none. Among us are people of great intellectual ability; people of the highest moral and ethical attainment; people of great tenacity and courage. The challenge is to put our talent, morality and courage to the right purposes, in the right direction.

I am positive that we will stand among the world's most powerful nations. The question is, whether after attaining that power can we set an example of how it can be used to build a better world, a world of justice and equity – justice and equity not for a few but for all?

Having said all this I am also very conscious that for me, like others of my generation, this is no longer our challenge. Alas, we are no longer young! The India of the future belongs to the young Indians and I believe that for all of them there is but one choice – that choice is to put all their abilities and energies into building the India of all our dreams.

If they are to do so – and I am ever optimistic that they shall – they have to recognise one thing: that great change takes place through small, sometimes invisible, steps. When Tribhuvandas Patel became Chairman of the Kaira District Cooperative Milk Producers' Union, little did he dream that it would one day become such a major player in India's food business. He was merely concerned with removing the obstacles from the district's milk producers.

Our dreams were enormous, our mission mighty. But we moved forward gradually, step by step. We knew we must begin by building a small, strong foundation towards change. We knew that this first foundation we built must be our very own. For as Gandhiji said: 'Be the change you want to see.'

POSTSCRIPT

TWICE, WHEN MY TERM AS CHAIRMAN, NDDB WAS ENDING AND I WANTED to retire, I was persuaded to continue. Ultimately, in 1998, my resignation was accepted and I retired from NDDB on 26 November.

From that very day I decided to work for Kaira's (now Kheda) dairy farmers; working for the cooperatives was an act of faith for me. It continues to be so. As I see it, faith is belief without reason. For those who believe, no explanation is necessary; for those who do not, no explanation is possible. I was fully convinced that I had imparted this deep and abiding faith in cooperatives to my colleagues at NDDB. Sadly, I now discover that it is not entirely so. I learn that the NDDB of today seems to be on a distinctly different path – that of building its own empire instead of helping cooperative milk producers, which was its single-point agenda!

I could have never imagined that those whom I had groomed, through the decades and who had been victorious through many a hard battle, would so quickly embark on a mission that completely undermines the very foundations, nay the very soul of the cooperative movement, posing a challenge to the cooperative movement itself. What is worse, all this is being done by using the very resources generated, through the years, for cooperative dairy development.

It must be kept in mind that NDDB is a Government of India

organisation, created by an Act of Parliament, with a mandate to follow the 'cooperative strategy'. NDDB was formed in 1965 to replicate the 'Anand pattern' of dairy cooperatives. Following this pattern essentially meant that production, processing and marketing would be firmly in the hands of milk producers. Throughout my time at NDDB, I was never tired of emphasising the importance of marketing, for it was clear to me that without absolute control over marketing, the farmers would not have any say in determining the price of the producers' milk and its quality. What NDDB has recently attempted goes completely against the principles and spirit of the Anand pattern of cooperatives.

NDDB has registered a company called Mother Dairy Fruits and Vegetable Private Ltd (MDF&VL), and has formed several other companies under it which are in competition with the State Dairy Cooperative Federations in the milk business. NDDB is now advocating joint ventures (JVs) with State Dairy Cooperative Federations. Under the JVs, Mother Dairy Foods Ltd will now hold 51 per cent equity while the State Dairy Cooperative Federation will hold just 49 per cent. MDF&VL will accept only a fixed quantity of milk and milk products and the state federations will not be allowed to sell their products to a third party. And as the final nail in the coffin, MDF&VL will handle the entire marketing exclusively. It seems to me that the dairy farmers have lost the most important component – that of marketing – and now have only production and processing (that too, limited by Mother Dairy, Delhi) in their hands. Will this not deprive the farmers of their due share of the consumers' rupee? The question then arises: how has the government permitted NDDB to form MDF&VL to compete with dairy cooperatives? The only explanation, I can think of, is that perhaps fancy words like 'liberalisation', 'globalisation' and others coined by powerful capitalist nations to penetrate the vast markets of developing nations, have blinded the vision of a few of our policy makers.

Thus far, my passionate objections to this major and alarming shift in NDDB's policies have not yielded the desired results. But I am a born fighter and I am determined to continue my fight until the end for the just cause of the cooperative movement of the dairy farmers of India. For what is at stake here is the future well-being of millions of small and marginal dairy farmers and landless labourers of India who depend on the daily income from the milk they supply to their cooperatives. I cannot give up hope that the present NDDB management will realise their mistakes and re-dedicate their energy and resources for the development of cooperatives, by following the 'cooperative movement strategy' which has proved time and time again, over these past five decades, to be the best suited for India's unique conditions.

It was only last year that I said in an article:

> When farmers are helped to become self-reliant, to manage their own affairs and to manage their livelihoods, they become the strong spine of the nation's democracy. Therefore it becomes the prime responsibility of the government to nurture (the) cooperative.
>
> I have fought against the efforts to undermine the interests of our farmers by vested interests – be they those of unscrupulous politicians, bureaucrats, businessmen or institutions – for all my life, and I will continue to do so unless someone shows me a better way of serving our nation's producers to become productive members of our society.

Till today, I have not found a better way. My unfinished dream will only be accomplished when the farmers of India have a level playing field to compete with other forms of businesses. Nobody can deny that the government and government agencies like the NDDB have a critical role to play in achieving this goal. There are some who always try to find fault with the cooperatives. My question to them is: has democracy in India worked as it should have? If not, is it the fault of democracy as a system, or does the fault lie in us? Similarly, I believe, cooperatives have 'not' failed. In fact, in India, there are only a few genuine cooperatives, like Amul. Merely registering as a cooperative is not sufficient. The cooperative so registered must also follow, unbendingly, the cooperative principles. If cooperatives have failed, it is because they have, so far, not been given a level playing field and because they have not been run as true cooperatives.

Cooperatives are people's institutions. I have great faith in the people's power. What our government needs to do is to simply empower the people. Only that can turn our country into a super power as envisioned in 'Vision 2020'. As Chairman of the Viksit Bharat Foundation, inspired by the President of India, our programmes are directed towards making India a truly developed nation.

Bureaucracies do not exist only in government. Tragically, most institutions, given time and allowed to grow big, tend to get bureaucratised. When employees begin to believe that the institution exists for them, rather than that they exist for the purpose and ideals for which the institution was built, then clearly, that institution has mutated into a bureaucracy. The revolutionary in me would demand that such an institution be broken down unabashedly, and built again anew.

Recently, there was an attempt by a coterie of vested interests to usurp control of the Institute of Rural Management, Anand, which I had set up in 1979 and unceasingly nurtured all these years. I tried to develop it as a national level institute that would serve the managerial needs of our rural people's institutions which, even decades after independence, have remained neglected. Serious attempts were made to remove me from the chairmanship of IRMA unconstitutionally, by maligning me through misuse of the media, and using the then Director of IRMA – whom I had removed due to serious misconduct. These were attempts to satisfy the coterie's own interests and convert the institute into a profit-making, private-management institution, like many others in the country, which cater to merely meeting the demands of private sector, particularly the corporates.

The people behind these attempts were people with considerable resources at their disposal. However, with appropriate and timely actions and with the help of those committed to IRMA's cause, I could keep IRMA from being derailed from its core mission – that of producing managers for India's under-managed rural organisations which believe in people-centred, equitable and sustainable development.

People may call me cold-blooded but I am very firm in setting correct precedents. This is a practice I will follow until the very end. As in life, so too in death. And because some things cannot be left to the people one leaves behind, I have already asked, for instance, that my body be cremated here in Anand. No special place and no special functions for me. As the poet, Alfred Tennyson, so sensibly said:

> Sunset and evening star
> And one clear call for me!
> And may there be no moaning at the bar
> When I put out to sea.

As I write these last few lines of my memoirs, my faith in the farmers of our country remains unshaken. Although I do feel betrayed by some whom I trusted, I continue to get support from unexpected quarters. The journey I began in Anand in 1949 still continues. I believe it will continue until we succeed .... Until India's farmers succeed.

31 December 2004

ANNEXURE

AMUL STATISTICS

	1953-54	2003-04
Turnover	Rs 55.47 lakh	Rs 54,088 lakh
Price difference	0.32 lakh	1,235 lakh
Gross profit	8.78 lakh	10,698 lakh
Total mandlies (societies)	64 nos.	1,059 nos.
No. of members	14,441	598,707
Milk procurement	108.63 lakh kg	2,539 lakh kg
Milk procurement price	Rs 7.00 per kg	Rs 200.00 / kg
Artificial insemination (AI)	1,673 nos.	675,707 nos.
AI centres	6 nos.	9,15 nos.
Pregnancy diagnosis'	479 nos.	245,967 nos.
Products	Ghee. Butter, Milk Powder, Casein	Ghee. Butter, Milk Powder. Cheese, Flavoured Milk, Cattle Feed, Chocolates, Nutramul, Baby Food, Ready to Eat Food, Bread Spread
		Installed chilling centres at Cambay, Kapdwanj, Balasinor and Khatraj
		Installed village chilling units at various societies
		Mogar Complex consisting of Chocolate Plant, Malted Milk Food Plant, Ready to Eat Food Plant, Bread Spread Plant
		Cattle Feed Plant
		Khatraj Satellite Dairy manufacturing Cheese varieties
		Amul Satellite Dairy at Pune

GUJARAT COOPERATIVE MILK MARKETING FEDERATION LTD

The Gujarat Cooperative Milk Marketing Federation Ltd (GCMMF) is India's largest food products marketing organisation. It is a state level apex body of milk cooperatives in Gujarat, which aims to ensure that farmers get remunerative returns and consumers are provided quality products.

Year of establishment	1973
Members	12 District Cooperative Milk Producers' Unions (details in attached list)*
No. of primary village societies	11,400
No. of producer members	2.36 million
Milk handling capacity per day	6.8 million litres per day
Milk collection (2003-04)	1.81 billion litres
Milk collection (daily average 2003-04)	4.96 million litres
Milk drying capacity	510 MTS per day
Sales turnover – 2003-04	Rs 28,820 million (yearwise details in attached list)**
Name of CEOs	V. Kurien, CMD – 1973 to 1984, and as Honorary Chairman since then
	J.J. Baxi – MD – 1984 to 1994
	B. M. Vyas – MD – 1994 to ...
No. of employees	755
No. of sales offices	49

* LIST OF MEMBER UNIONS

1. Kaira District Cooperative Milk Producers' Union Ltd, Anand
2. Mehsana District Cooperative Milk Producers' Union Ltd, Mehsana
3. Sabarkantha District Cooperative Milk Producers' Union Ltd, Himatnagar
4. Banaskantha District Cooperative Milk Producers' Union Ltd, Palanpur

5. Surat District Cooperative Milk Producers' Union Ltd, Surat
6. Baroda District Cooperative Milk Producers' Union Ltd, Vadodara
7. Panchmahal District Cooperative Milk Producers' Union Ltd, Godhra
8. Valsad District Cooperative Milk Producers' Union Ltd, Valsad
9. Bharuch District Cooperative Milk Producers' Union Ltd, Bharuch
10. Ahmedabad District Cooperative Milk Producers' Union Ltd, Ahmedabad
11. Rajkot District Cooperative Milk Producers' Union Ltd, Rajkot
12. Gandhinagar District Cooperative Milk Producers' Union Ltd, Gandhinagar

AFFILIATED DAIRIES

1. Satellite Diary, Mahemdavad
2. Vidya Dairy – Students' Dairy at Anand
3. Mother Dairy, Gandhinagar

** SALES TURNOVER 1993-94 TO 2002-03

Sales Turnover	Rs in million
1993-94	9,889
1994-95	11,140
1995-96	13,793
1996-97	15,537
1997-98	18,836
1998-99	22,192
1999-2000	22,185
2000-01	22,588
2001-02	23,365
2002-03	27,457

TARGET, ACHIEVEMENT AND FUNDING DURING OPERATION FLOOD PROGRAMME

Phase-I

Target To set up dairy cooperatives (DCSs) in rural areas (in Karnataka, Rajasthan and Madhya Pradesh), provide physical and institutional infrastructure for milk procurement, processing, marketing and production enhancement services at union level, establish city dairies and link city dairies with rural cooperative societies for milk supply.

Achievement DCSs – 13,270 (1,588 in 1970-71)
Membership – 17.47 lakh (2.78 lakh in 1970-71)
Milk procurement – 25.60 lkgpd (5.20 lkgpd in 1970-71)

Funding Funds of Operation Flood-I were generated by the sale of SMP (126,000 tons) and BO (42,000 tons). A total of Rs 116.54 crore was invested in the implementation of the programme.

Phase-II

Target Successful replication of Anand Pattern – a three tier cooperative structure of societies, unions and federations already laid down during Operation Flood-I.

Achievement DCSs – 34,523 (1984-85)
Membership – 36,32 lakh
Milk procurement – 57.8 lkgpd

Funding Total outlay was Rs 273 crore (World Bank – USD 150 million and balance through EEC – commodity assistance).

Phase-III

Target Strengthening institutional management, productivity enhancement, consolidating National Milk Grid.

Achievement Milkshed – 170

PHYSICAL TARGETS AND ACHIEVEMENTS OF OPERATION FLOOD

Particulars	Operation Flood-I July 1970-Mar 1981		Operation Flood-II Apr 1981-Mar 1985		Operation Flood-III Apr 1987-Mar 1992	
	Target	Achievement	Target	Achievement	Target	Achievement
No. of milksheds covered	18	39	155	136	190	170
No. of village dairy societies organised	–	13,270	29,000	34,523	70,000	64,494
Farmer families covered (lakh)	10	17.47	34.8	36.3	67	79.3
Rural milk procurement (LLPD*)						
Average	–	25.6	56.9	57.9	137	93.9
Peak	–	33.9	–	79	183	113.4
Urban milk marketing (LLPD)	–	27.8	43	50.1	103	83
Rural dairy processing capacity (LLPD)	29.8	35.8	76	87.8	200	152.2

* LLPD: lakh litre per day

NATIONAL MILK PROCUREMENT BY OPERATION FLOOD, OTHER ORGANISED AND INFORMAL SECTORS (PER CENT)

	1961	1972	1980/81	1988/89	1992/93	1994/95	1995/96	1996/97
1. Total production of milk (MMT*)	10.40	23.00	31.60	53.70	59.00	64.00	66.10	70.10
2. Organised sector	0.80 3.68%	1.30 5.55%	3.10 9.93%	5.80 10.88%	n.a. n.a.	6.40 10.10%	n.a. n.a.	n.a. n.a.
3. Operation Flood cooperatives	n.a.	0.24 1.03%	1.01 3.21%	3.58 6.67%	3.85 6.53%	3.75 5.85%	4.00 6.05%	4.40[a] 6.28%
4. Non-Flood organised sector		1.04 4.52%	2.12 6.72%	2.46 4.20%		2.63 4.16%		
5. Handled by Operation Flood cooperatives as percentage of organised sector	10.56%		32.33%	61.33%		58.77%		
6. Handled by the information sector	19.65 26.30%	21.72 94.44%	28.46 90.07%	47.87 89.12%		56.72 89.88%		

* MMT: million metric tons

DAIRY COOPERATIVES (DCSs) – 72.744
MEMBERSHIP – 93.14 LAKH

Funding

World Bank – USD 360 million
EEC – Rs 222.6 crore (75,000 tons of SMP and 25,000 tons of butter oil)
NDDB – Rs 206.3 crore

Item	1971	Phase-I 1981	Phase-II 1985	Phase-III 1995	Target 1996
No. of milksheds	5	39	136	170	170
No. of DCS ('ooo)	1.6	13.3	34.5	72.7	70
No. of farmer members (lakh)	2.8	17.5	36.3	93.1	*
Milk procurement (lkgpd)	5.2	25.6	57.8	109.4	115
Processing capacity					
Rural dairies (llpd)	6.8	35.9	87.8	194.0	193.7
Metro dairies (llpd)	10.0	29.0	35.0	72.0	72.40
Milk drying capacity (tpd)	Na	261.0	507.5	842.0**	974
Milk powder production ('000 ton/annum)	22.4	76.5	102.0	195.0**	*
AI centres ('000)	NA	4.9	7.5	16.28**	16.50
AI done (lakh)	NA	8.2	13.3	37.9**	39.5
Cattle-feed capacity ('000 tons/annum)	NA	1.7	3.3	4.9**	5.0
Investment (Rs crore)	NA	116.54	277.17	896.21	1,303.10

* Target already achieved.
** Till end 1995.

INDEX

Aarey Milk Colony, Bombay, 17
Adulteration, 171
Advani, L.K., 219, 221–22
Advertising, 55, 69, 71, 75, 137
Advertising and Sales Promotion Company (ASP), 75
Agriculture, 208
Agriculture College, Pune, 7
Agriculture Ministry, Government of India, 22, 26–27, 101, 127–28, 135, 136, 153, 171–73
Agro-climatic conditions, 139, 189
Ajarpura, Kaira, 96–97
Akbarallys, Bombay, 69
All Saint's C.S.I. Church, Trichur, 40
Alva, Margaret, 148–49
Alvares, Claude, 145
Amrit Kaur, Rajkumari, 91
Amul, *see* Kaira District Cooperative Milk Producers Union Limited (KDCMPUL), Anand
Amul butter, 55, 57, 63, 66, 68, 69, 74, 75
Anand, Anand pattern of cooperatives, 2, 9–23, 26–28, 30–31, 32, 40–42, 45, 47–48, 50–53, 57, 60, 63, 73, 76–78, 80–81, 83–85, 88, 91, 92, 94, 97, 98–102, 104–5, 107, 111, 113–4, 121, 123, 129, 132, 134, 140, 142, 148–49, 156, 158, 164, 167–68, 177, 179, 181, 187–88, 190, 192, 194–95, 198, 202, 204, 206, 210, 211, 213, 215, 221, 223, 225, 228, 233, 236

Anandalaya, 133
Anchor butter, 63
Aneja, R.P., 150
Animal husbandry, 139
Anthony, 20–21
Artificial insemination, 80, 136
Auditor General's Act, 156
Awards and recognition, 221–23

Baby food, 69, 71–72, 125, 128, 129
Balancing system, 138
Baroda Cooperative, 123
Barot Kaka, 19, 47
Beedi Tobacco Research Station, 20
Benegal, Shyam, 132–34
Bhatty, I.Z., 152
Bhavnagar, Gujarat, 173
Birla House, 38
Birlas, 20, 38
Bombay: cattle population, 108–09; buffaloes, mortality, 109; state government, 45, 54
Bombay Milk Scheme (BMS), 13, 17, 29–30, 32, 58, 88, 104, 110, 128
Bombay Milk Scheme (BMS) Laboratory, Anand, 26
Branding, brand building, 54–55, 72, 137
Breastfeeding, 71
British rule, 14, 16, 204
Bureaucracy, 18, 21, 58, 84, 90, 101, 103, 104, 106, 113–15, 120, 124, 132, 135, 155, 159–60, 162, 164, 180–81, 197–98, 201, 205, 207–8, 219–20, 234
Butter oil, 112

Calcutta Milk Scheme, Calcutta, 104, 128
Calf mortality, 109
Capitalism, 56
Cast iron, 8
Caste system, 78
Cattle breeding and nutrition, 149
Cattle-Feed Compounding Factory, Kanjari, 94
Cattle-feed plants, 103
Central Food Technological Research Institute (CFTRI), Mysore, 69–71, 165
Chaklashi village, 16
Charles, Prince of Britain, 186
Chavan, Y.B., 32, 57
Chothani, A.A., 215
Colombo Plan, 34, 54
Commitment, 131, 144
Competition, 67–68, 71, 99, 103, 138, 208, 227–30
Comptroller and Auditor General of India (CAG), 156–58
Condensed milk, 59–60, 69, 128
Congress, 28, 32, 38, 46
Constitution of India, 157
Consumer acceptability, 171
Contractors' lobby, 85, 89
Cooperative movement, 131, 234
Cooperative Societies Act of India, 99
Corruption, 88, 128, 129, 157, 177, 192–93, 197, 200–1
Cotton Corporation of India, 207
Cow slaughter issue, 182–85

DaCunha, Sylvester, 75
Dairy cooperatives, 77, 188, 227–28
Dairy development, 116, 226
Dairy engineering, 6
Dairy equipment: import, 148; manufacturing industry, 153
Dairy industry, 16, 56, 94, 130, 111, 189

Dairying department of the British government, Bombay, 16
Dalaya, Harichand M., 7–8, 25, 30, 36–37, 40–41, 44–46, 48–50, 58, 70, 76, 88, 93, 107, 216
Dastur, N.N., 152
Dave, N.S., 88
Davies, T. Glen, 43–45
Delhi Milk Scheme (DMS), Delhi, 84–91, 101, 104, 122, 128
Democracy, 76–77, 82, 204–5, 234–35
Denmark: dairy farmers, 66
Desai, Dinkarrao, 44–45, 47, 62
Desai, Morarji, 15–17, 38, 50–51, 53, 92, 93, 135–36, 167, 183
Deshmukh, C.D., 33
Devakaran Nangi Trust, 69
Development, development process, 58, 80, 82, 139, 162–63, 199, 228
Devi Lal, 179
Dhar, D.P., 155
Dhara, 170–72
Diacetyl, 65
Distribution network, 137
Dudani, A.T., 7, 161
Dutch Volma milk powder plant, 46
Dystokia, 79

Ecology, 81
European Economic Community (EEC), 112, 117–18, 135, 136, 154
Empire Stores, Delhi, 69
Environmental degradation, 81
European Union (EU), 117, 227
Evaluation Committee (Jha Committee), 151–54
Exploitation, 15–16

Fakhruddin Ali Ahmed, 142
Farm prices, 146

Farrall, *Prof.* A.W., 34
Federal system, 77
Fernandes, Eustace, 75
Fodder situation, 189
Food aid, 144–46
Food and Agriculture Organization (FAO), 107, 115, 117, 131
Food Corporation of India (FCI), 207
Food security in milk, 107
Forbes, Darius, 50, 52
Ford Foundation, 107
Foreign investment, 61
Formalin, 31
Foster, 65, 67
Free market economy, 227
Freedom movement, 14
Fruits and vegetable cooperative, 207

Gagrat, 74
Gandhi, Indira, 52–53, 129, 151, 155, 165, 166–67, 206
Gandhi, M.K., 204, 231
Gandhi, Rajiv, 167, 174–76, 178, 181, 212, 225
Gandhi, Sonia, 177–78
George, Shanti, 145, 162
George, Vincent, 175
Ghandy, *Sir* Jehangir J., 5
Glaxo, 70–72, 128
Globalisation, 227, 233
Golwalkar, M.S., 182–83, 185
Government research creamery, Anand, 18–19, 22
Grant Advertising, 55
Guindy College of Engineering, Madras, 3
Gujarat Agricultural University, 197, 214
Gujarat Cooperative Milk Marketing Federation (GCMMF), Anand, 1–2, 133, 138, 171, 214, 221–23
Gujarat Education Society, 133

Gujarat Electricity Board (GEB), 197–201, 214
Gujarat government, 73, 103
Gujarat Provincial Congress Committee, 16

Halse, Michael, 88, 107
Heredia, F.J., 48, 95–96
Heredity, 8
Hussain, 7, 9
Hussein, Injaz, 190–91

Illustrated Weekly of India, 145
Imdad Ali, 113
Imperial Dairy Research Institute, Bangalore, *see* National Dairy Research Institute of India
Import duty, 189
Import of milk and milk products, 29–30, 63–64, 125–29, 168, 189–90
Indian Companies Act, 1956, 121
Indian Council of Agricultural Research (ICAR), 180
Indian Dairy Corporation (IDC), 121, 126, 150–1, 152–54, 159, 160, 161, 163, 166
Indian Institutes of Management (IIMs), 210; Ahmedabad, 107, 211–12; Calcutta, 211
Institute of Rural Management, Anand (IRMA), 2, 133, 211–13, 221, 224, 235–36
Institute of Social Studies, Hague, 145
Integrity, 25, 27, 169, 215
International Dairy Congress, New Delhi, 1974, 217
International Labour Organization (ILO), 193

Jagjivan Ram, 155, 164–66, 178
James Wright, Calcutta, 69
Jha, L.K., 151–54
Jhakar, Balram, 177

Jinnah, Fatima, 91
Johnson, Lyndon B., 81
Jute Corporation of India, 207

Kaira, Gujarat: dairy industry,
    12–13; milk cooperatives, 23;
    milk strike, 17
Kaira Can, 76
Kaira District Congress Committee,
    15
Kaira District Cooperative Milk
    Producers Union Limited
    (KDCMPUL), Anand, 1–2, 15,
    17, 18, 23, 24, 27, 28, 32, 33,
    35–36, 41, 43–47, 53, 55–58,
    60, 66, 67, 71, 73, 75–81, 83,
    88, 91, 98–105, 113, 121–23,
    125, 129, 131, 134, 137–38,
    140, 158, 167, 170–72,
    178–80, 193–94, 197, 200,
    205, 208, 213, 214, 216, 221,
    223, 231, 235
Kantawala, Jit, 55
Kanvide, 133
Kapadia, Harshad, 76
Karamsad village, 14
Kasturbhai Lalbhai, 70
Katrak, Usha, 75
Kay, *Prof.* H.D., 35
Kennedy, John F., 81
Kesteven, 117
Khadi saris, 38
Khurody, Dara, 17, 29–32,
    42–45, 47, 49, 51, 52, 57–58,
    62
Kiriya Milk Industries, Sri Lanka,
    192–94
Kodandapani, 7, 10, 19
Kosygin, Alexei, 186–88, 201
Kothavala, Zal R., 11, 68–69
Kreeber, 59–61
Krishnamachari, T.T., 63–64, 69
Kumaratunga, Chandrika,
    192–94
Kurien, K., 75

Kurien, Puthenparakkal, 3

Labour laws, 197
Larsen and Toubro (L&T), 25, 44,
    46, 49; Niro powder plant, 46;
    Silkeborg Pasteuriser, 25
Liberalisation, 173, 208, 227, 233
Liquid milk schemes, 116
Literacy, 3
Loyola College, Madras, 3

Maharashtra: sugar cooperatives,
    205–6
Makwana, Yogendra, 150
Malnutrition, 147
Manmohan Singh, *Dr.*, 208
Mansoor, 7, 9
Manthan, 133–34
Marketing, 54–56, 63, 71,
    122–25, 136, 137, 140–41,
    144, 153, 169, 171–72, 174
Marshall, J.N., 50–51
Matthai, Cherian, 4
Matthai, John, 4–5, 9, 11
Matthai, *Mrs.* John, 40
Matthai, Ravi, 211–12
McNamara, Robert, 149
Medora, Pheroze, 7, 26–27,
    30–31, 73
Mehsana Dairy Cooperative,
    137–38
Mehta, Balwantrai, 95
Mehta, Jivaraj, 43
Metal Box, 76
Michigan State University, 7, 9,
    33–34, 116
Milch animals,
    slaughter/destruction, 108–11
Milk per capita availability and
    consumption, 108, 226
Milk collection, 77–79, 138
Milk Commissioner, Bombay, 14,
    16, 30, 43
Milk cooperatives, 14, 17, 24, 106,
    111, 123, 133, 136–37

Milk industry, British government's intervention, 12
Milk, nutritional components, 147
Milk powder, 51, 57, 69, 71–74, 112, 126–29, 135, 136–37, 190; from buffalo milk, 21, 35, 42, 45, 48, 52, 59, 72, 189; from cow's milk, 35, 72
Milk procurement process, 89, 122
Milk production, 108, 152–53, 226, 231; decline, 111
Milk rationing, 85, 89–90, 141
Mitra, Ashok, 182
Modernisation, 82
Moga, Nestle milk plant, 61
Mother Dairy Foods Ltd, 233
Mother Dairy Fruits and Vegetable Private Ltd (MDF&VL), 233
Mother Dairy, Delhi, 140, 142–43, 161, 233
Multinationals (MNCs), 28, 59–60, 194–95
Munshi, K.M., 33

Naar, 62
National Cooperative Dairy Federation of India Ltd (NCDFI), 2
National Dairy Development Board (NDDB), Anand, 2, 69, 76, 101–5, 108, 111, 116, 117, 119–22, 126–34, 136, 139, 145, 146, 148–54, 156, 158–63, 165–66, 168–70, 172–75, 178, 181, 188, 192, 203, 204, 205, 206–8, 216–17, 218–21, 224–26, 232–34; Act 1987, 154, 156, 158, 202, 219; autonomy, 158–59
National Dairy Research Institute of India (NDRI), Bangalore, 6, 10, 11, 18–19, 161
National Research Development Corporation (NRDC), 70–71
National Tree Growers' Cooperatives, 207
Nationalism, 183
Nazir Ahmed, 117, 120
Nehru, B.K., 151
Nehru, Jawaharlal, 37–38, 47, 51–53, 92, 167
Nestle Alimentana, Vevey, Switzerland, 58–61
Nestle India, 59, 61, 172, 193–95
Netherlands, Queen of, 148–49
New Zealand: aid and assistance for Kaira milk cooperative, 35, 53–54
New Zealand Cooperative Dairy Company, 34
New Zealand Dairy Board, 63, 126–27, 193, 195
New Zealand Dairy Research Institute, Massey Agricultural College, 34–35
Non-governmental organizations (NGO) sector, 211, 213

Oilseed Growers Cooperatives, 169–70
Oilseed Growers Unions, 170
Oilseeds, 168, 174
Operation Flood, 106–8, 111–15, 116–39, 140–46, 147–55, 161–64, 186, 210, 212
Operation Flood-III, 154
Oxfam, 94

Pandya, *Prof.*, 24
Pansora village, 31
Parpia, H.A.B., 182
Parsi, 65
Parthasarthy, Vibha, 133
Partition, 37
Pasteurisation, 6, 56, 141
Patel, Amrita, 217, 219–20, 224–25
Patel, B.R., 115, 117
Patel, Babubhai Jessubhai, 43
Patel, Chimanbhai, 39, 179, 201

Patel, H.M., 46, 167–68, 217, 220
Patel, Maganbhai D., 24–25
Patel, Maniben, 37–38, 41–42, 47–48, 91
Patel, Ramanbhai Punjabhai, 96–97
Patel, Sardar Vallabhbhai, 14–18, 37–39, 47–48, 54, 58, 82, 94,100, 107, 167, 204
Patel, Tribhuvandas, 15, 17–18, 23–25, 27–28, 32, 36–38, 41–41, 43, 55, 58, 76, 80, 99, 131, 133, 162, 167, 213–16, 218, 231
Patriotism, 130
Pestonjee. Edulji, 11–13, 63, 65–69
Pestonjee, Minoo, 68–69, 74
Petersen, Axil, 25, 44
Philip, K.M., 54–55
Pitroda, Sam, 225
Planning process, 180
Politics, politicians, 89, 155, 162, 180–81, 205, 208, 214, 218, 234
Polson, 13–14,15, 17, 42, 57, 62–69, 74
Poverty, 81, 147, 230
Power cooperatives, 200
Power theft, 201
Prasad, John, 143
Prasad, Rajendra,18, 47
Press Council of India, 146
Press Syndicate, 55
Private sector, 12, 56, 138, 173, 209

Quality, 66

Racism, 8
Radeus Advertising, 132–33
Ramachandran, V., 166
Ramon Magsaysay Award for Community Leadership, 1963, 221

Rann of Kutch: salt workers, 202–4
Rao, Bala, 20
Rao, P.V. Narasimha, 191–92
Rashtriya Swayamsewak Sangh (RSS), 182, 185
Rau, S.K., 152
Ridett, *Prof.* William, 35, 42
Royal Commission of Agriculture, 125
Rural economy, 124, 137
Rural employment, 227

Sabin, Donald, 43–45
Sadhwani, H.T., 90–91
Sagar, 138
Sahay, Vishnu, 33–34
Samanwaya Puraskar, 221
Sarabhai, Vikram, 155, 211, 225
Sarkar, *Justice*, 181–82
Sathwamithranand, Swami, 221
Scott, *Prof.* Kelvin, 34
Seshan, T.N., 176
Shah, Manubhai, 58, 60
Shankaracharya of Puri, 182–83
Shastri, Lal Bahadur, 94–104, 129
Shivaraman, 72–73, 115
Sikka, L.C., 87, 88
Singh, L.P., 113–14
Singh, Rao Birendra, 149–51, 154–55, 206
Singh, T.P., 159, 160
Social and economic change, 80–1, 101
Social fabric, 78
Societies Registration Act, 1860, 116
Spencer and Co., Madras, 69, 137
Sri Lanka: Anand pattern of milk cooperatives, 192–95
State Cooperative Dairy Federations, 233
State Trading Corporation (STC), 173

Subramaniam (of CFTRI), 70
Subramaniam, C., 84, 88–90, 101, 102, 220
Subsidies, 89
Sugar cooperative sector, 205–6
Surat, 50
Swaminathan, M.S., 165
Syrian Christian community, 3, 4

TTK & Sons, 69
Taj Hotel, Bombay, 21
Tata Group, 6, 130
Tata Industries, 5–6, 7, 9, 10
Tata Iron and Steel Company (TISCO), 4–7, 82, 143
Tata, J.R.D., 82, 130
Teddington Chemicals, Andheri, Bombay, 44–45
Telia rajahs, 168, 169–70, 176
Tendulkar, Vijay, 134
Tennyson, Alfred, 238
Thimmaiah, *Captain* K.S., 4
Thompson, J. Walter, 55
Token acceptance mechanism, 143
Tondamman, 193
Toubro, 50, 53
Transformation, 213
Tree growers' cooperative, 207
Tribhuvandas Foundation, 79
Trichur, 39–40

Union Carbide, 9
United Kingdom National Institute of Research and Dairying, Reading, 35
United Nations (UN), 120
United Nations Inter-Agency Mission, 144
United Nations International Children's Education Fund (UNICEF), 42–46, 53, 143; Food Conservation Division, New York, USA, 43
United States of America, National Rural Electric Cooperative Association, 200
University Training Corps (UTC), 3–4; Madras, 4

Vajpayee, Atal Behari, 218
Variava, R.H., 41–42, 66–67, 69, 74
Variava, Zarine, 41–42
Vegetable oil (edible oils), 168–69, 173, 186–87, 206; import, 168–69, 171, 173; procuring, processing and marketing, 168
Vending machine, 142–44
Vending system, 153
Venkataramanan, R., 88
Verghese Molly (*nee* Peter Susan), 39–42, 49, 52–53, 70, 91–92, 218, 223, 224
Verghese, Nirmala, 91–92, 223
Verghese, P.K., 3
Veterinarians, 78
Veterinary health care, 136–37
Viksit Bharat Foundation, 235
Village cooperatives, 76–78
Vishnu Bhagwan, 151
Vishwa Hindu Parishad (VHP), 221
Voltas, 137

Ward, Barbara, 81–82
Women empowerment, 78
World Bank, 107, 115, 136–37, 149, 154, 188
World Food Prize, 1989, 222
World Food Programme (WFP), 116–17, 120, 135
World War I, 18, 24

ACKNOWLEDGEMENTS

MANY PEOPLE HAVE BEEN INVOLVED IN THE MAKING OF THIS BOOK. I WOULD like to begin by thanking Tom Carter who helped ascertain the many facts and figures for the book. K. Kurien's valuable suggestions provided the right stimulus and prompted us ahead. Jacob Matthai helped me immensely in recollecting some of the incidents I had forgotten. I am grateful to Roger C.B. Pereira who coordinated and provided valuable inputs.

P.A. Joseph, my Executive Assistant, was a moving force, helped me collect a lot of information. A special thanks is due to him. And above all I would like to thank Gouri Salvi for putting my thoughts into words.